The New Quantum Era

First Edition

An Outsider's Introduction

Sebastian Hassinger

<packt>

The New Quantum Era

First Edition

Portfolio Director: Kunal Chaudhary
Relationship Lead: Denim Pinto
Program Manager: K. Loganathan
Content Engineer: Sushma Reddy
Technical Editor: Aditya Bharadwaj
Indexer: Manju Arasan
Production Designer: Shantanu Zagade
Growth Lead: Vinishka Kalra

First published: May 2026

Production reference: 1150526

Published by Packt Publishing Ltd.
Grosvenor House
11 St Paul's Square
Birmingham
B3 1RB, UK.

ISBN 978-1-80778-737-0

www.packtpub.com

This book is the product of many fascinating conversations with brilliant, patient scientists without whom I wouldn't understand any of this stuff. Thanks to Pat Gumann, Chris Lirakis, Dimitri Maslov, John Preskill, Elise Crull, Grant Salton, Peter Shor, Scott Aaronson, Sebastian Weidt, Charles Bennett, Prineha Narang, Yasunobu Nakamura, Dave Bacon, Ken Brown, Brian DeMarco, Susannah Glickman, Lin Lin, and so many others.

Special thanks to those who read early drafts and provided invaluable feedback: Gilad Ben-Shach, Pedro Lopes, Gabe Paré, Mike Loukides and Emily Edwards.

My infinite adoration and appreciation for Ayann, whose beautiful mind keeps me endlessly stimulated and curious, and whose love and depthless confidence in me kept me buoyed from the very beginning of everything.

~ Sebastian Hassinger

Foreword

Quantum computing is the most serious scientific bet humanity has made about the future of computation – built on the deepest understanding of physics we possess. Moore's Law, the engine of computational progress that has driven economic growth and scientific capability for six decades, may be approaching fundamental physical limits. And the great challenges facing humanity in this century – in medicine, materials research, and AI – demand computational power that classical architectures alone cannot provide.

And yet the quantum computing story remains extraordinarily difficult to tell well. There is a version that gets repeated too often: a revolution is imminent, machines of unimaginable power are just around the corner, and every encryption scheme, every unsolved optimization problem, every molecular simulation is about to yield. As has happened before with every technology that outpaces the public's ability to understand it, the real progress gets buried under the noise of overstatement.

There is an equally unhelpful counter-reaction: the weary shrug of the technically literate, whose skepticism, however well-earned, underestimates the inexorable power of scientific understanding to advance beyond what seemed impossible a generation before. In that telling, quantum computing has been ten years away since 1994, and will stay that way.

What is missing in both accounts is genuine understanding – not a simplified explanation that makes you feel informed without actually being so, but real explanatory knowledge of what is happening, why it matters, and what remains deeply uncertain.

That is what this book provides.

I have known Sebastian Hassinger for years. He has been in the room – at the conferences, in the conversations with researchers and founders – and not merely as an observer. He has helped shape the way this field understands and communicates itself. What has always distinguished his thinking is a refusal to be satisfied by approximation. He earned his understanding of quantum computing through years of sustained engagement with the world's leading researchers and a willingness to keep asking the foundational questions that specialists assume are already answered. That combination is exactly what makes this book unlike anything else in the field.

Running The Quantum Insider, I spend my days inside this ecosystem, tracking the capital, following the research, watching national programs take shape. I live and breathe this industry. The way he traces how the quantum market actually came together – the people, the decisions, the pivotal moments that never made the public record – is a story that until now has only existed in the memories of those who were there. It taught me a great deal.

I firmly believe that we must develop quantum technology if humanity is to achieve its most ambitious goals. Making good decisions about quantum computing – as investors, as technologists, as policymakers, as citizens – requires the kind of understanding this book builds. Not blind faith in the hype, and not reflexive dismissal of it. Something harder and more useful: an honest reckoning with what is known, what is not, and what the history of building entirely new categories of technology suggests about the road ahead.

Sebastian has written that book. It deserves your attention.

Alex Challans CEO, Resonance (The Quantum Insider)

Contributors

About the author

Sebastian Hassinger is a veteran technologist with decades of experience in emerging technology markets, including the early commercialization of the internet. He has held key roles at Apple and IBM, contributing to internet infrastructure, mobile computing, and cloud technologies, and holds multiple patents. Since 2017, he has worked in the quantum computing industry, first with IBM, then with Amazon Web Services and now as an independent consultant and advisor. He is also the co-host of The New Quantum Era podcast, where he interviews leading researchers and engineers shaping the future of quantum technology.

About the reviewers

Gilad Ben-Shach is the VP of Business Development at Quantum Machines. He previously worked at IBM Quantum, leading the Commercial Ecosystem Program and Product Management. He completed his PhD in condensed matter theory at Harvard.

Emily Edwards is an expert in science communication with a focus on developing and studying multimedia content to increase awareness, interest, and knowledge about quantum information science. Since 2011, she has co-led the creation of multiple resources for the public of all ages to learn about this complex, yet critical area research and technology. This includes the online glossary "The Quantum Atlas," and a quantum computing game world for middle grade students. She received her PhD from the University of Maryland in the area of ultracold atomic physics and was a postdoctoral researcher in ion trap quantum information science. Edwards is currently an Associate Research Professor at Duke University and was elected a Fellow of the American Physical Society in 2022.

Dr. Pedro LS Lopes is a researcher and educator in quantum computing. He is an expert in quantum matter and devices, and quantum information theory, with a career spanning top research institutions in the USA and Canada. He has operated and managed multi-institutional scientific programs at the interface of private industry, academia, and national laboratories. With his expertise, he has played advisory roles in quantum technologies for organizations such as NATO and GESDA. He runs an education and scientist relations team at QuEra Computing, is the author of over 30 scientific works, and runs the blog on science management "Set Physics to Stun."

Gabriel Paré works as a software engineer developing design automation tools for the manufacturingindustry. He has a passion for cracking hard problems and teaching others to do the same. He is hopeful about all the future brings despite his library of science fiction.

Table of Contents

Preface

Technologies are tools that human beings craft out of hard-won understanding of some aspect of our world, what British physicist David Deutsch calls explanatory knowledge. These tools have allowed us to attain capabilities of both a creative and a destructive nature unimaginable for any other form of life with which we are familiar. Explanatory knowledge has the power to reshape our existence in ways that convey maximum benefits to us, or to utterly destroy us. This terrible, marvelous gift is our central narrative. It has provided abundance to the point of utter luxury, but the costs incurred over the centuries are coming due. It seems entirely likely that this century will see a singular moment in which our fate will be determined. The weight of our debts will overwhelm us, exhausting our world and dooming us to whatever horrors lie beyond complete environmental collapse, or our toolmaking will surge ahead, maintaining the Anthropocene with a surplus available to service and eventually pay off our debt.

The scientific age gave us the crude tools of combustion and electricity to power the engines of the industrial era, and the first quantum age gave us the tools of the post-industrial information age, where we reside today. This digital tool set may be enough to win our struggle against the destructive externalities of the machine of humanity, but it's not a foregone conclusion. In fact, the probability of our success is unknowable at this point in time, only to be revealed after that decisive moment coming in the future. Either we develop and use the tools to sustain ourselves, or we don't. One hundred percent, or zero.

In anticipation of that fulcrum point, we are collectively in the process of mobilizing an historic, towering collection of resources with which we hope to stack the deck in our favor. The Inflation Reduction Act of 2022 directed hundreds of billions of dollars in subsidies toward the decarbonization of the US economy, in many cases with no effective cap — the legislative equivalent of "whatever it takes." Technologies such as solar panels, already the cheapest way to generate electricity, will play a vital, irreplaceable role.

However, our current technologies are rooted in the extractive, inefficient techniques gleaned from 19th century scientific understanding. The chips that power our most powerful applications are power hungry and wasteful, and the same is true for all of our technologies, especially our most advanced ones. As we debate whether today's machine learning software will ever achieve a level of artificial general intelligence, or ability to generate explanatory knowledge, comparable to human beings, it's worth noting that a single query to a frontier AI model can consume anywhere from a fraction of a watt-hour to tens of watt-hours of electricity, depending on the complexity of the task. The human brain, by contrast, runs on roughly 20 watts — continuously — while simultaneously managing perception, movement, memory, and everything we call thought.

In fact, throughout nature we have found processes that do work at efficiencies that are unimaginable with our current explanatory knowledge. Plants use photosynthesis to create energy with only beneficial byproducts. Bacteria can fix nitrogen and eat plastic with no hazardous wastes. Science is slowly building our knowledge of these processes, in the hopes we can learn how to make abundance for ourselves without continuing to destroy our planet. In all cases, these natural processes at some level depend on quantum mechanics to work their magic.

The theory of quantum physics represents our best explanation of the operating system of our universe, and in fact underpins the major components of our current digital age. Semiconductors used for microprocessors work because we understand and engineer their structures at a nanoscale where quantum mechanics describes behaviors of materials. Lasers were invented because of our understanding of photons and their quantum mechanical characteristics. However, quantum physics' contribution is indirect. The material we make microprocessors with provides a substrate on which electrical circuits are drawn, and the work is done at a scale that is orders of magnitude larger than the subatomic. Billions of electrons flowing through conductive leads and billions of photons streaming along a fiber optic strand behave in much the same way as those flowing through a light switch or from a light bulb in your home, which is to say, classically, not taking advantage any of the unique behaviors of quantum mechanics.

This has been our state of affairs, but now, in the 21st century, we are seeing the dawn of a new quantum era. Decades of effort in the field of quantum mechanics has begun to translate into an entirely new set of engineering techniques which are being used to invent new technologies. These technologies, such as quantum sensing, quantum communication, and quantum computing use quantum behaviors directly, leveraging them in the service of the desired application.

In December of 2017, I attended the Think Q Summit at IBM's T.J. Watson Lab, in Westchester County, New York. I worked for IBM at the time, though I was not affiliated with the Research division. In fact, at that point I had no idea what quantum computing was, what it might be used for, or just how mature it was — or was not. I was used to very early stage emerging technologies so I suppose I expected something like mobile computing in the nineties, like a Motorola Envoy portable device in 1994, with 1 megabyte of RAM, a 4,800 bit per second dial up modem, and weighing almost 2 pounds. A tantalizing proof-of-concept with a path to usefulness measured in years. The story of quantum computing that IBM presented at the summit was something else entirely.

Instead, on the auditorium stage, physicists were talking about future machines that beginning to use quantum mechanical behaviors to carry out computation. Instead of marketing fluff, we were served up complex linear algebra formulae. In place of a product roadmap, we got a tour through the way IBM was making chips which manifested "artificial atoms" that, when poked and prodded with microwave signals, would do math. Instead of applications there were open ended discussions about what they believed these machines might do, one day, if we got lucky. I found the most bleeding edge technology, closer to fundamental scientific research than anything I'd seen before or will ever likely see again. I was hooked.

A few months later I was part of the IBM Quantum team, and though I left the company in 2021, I still work in the quantum computing industry. I expect to finish my career working in this industry. Finding a way into this field I found so compelling was an intimidating challenge back in 2017. I am not a physicist, and my formal mathematics education ended with Calculus 2, which is the only class I ever failed, and I failed in an utterly miserable way. It's a long story but if I tell you I got nearly 100% in Cal 1 the first semester, you can infer that I had a very strange year as a freshman at University of British Columbia. So, with scant practical tools to understand the nitty gritty of the linear algebra of quantum algorithms, I had to find another way to build up a competency. I fell back on the tools that I knew from my liberal arts and humanities training. I read. A lot. I read a lot that I didn't understand, really, but I grasped at the conceptual frameworks the authors used to convey the meaning of their work. I asked a lot of questions, often dumb questions, of the talented and very patient researchers I worked with at IBM and various universities.

The fuzzy beginnings of the understanding of the science was augmented by my experience in emerging technology from the start of my professional career, establishing an Internet service provider in Vancouver the year the NSF's Acceptable Use Policy first allowed commercialization of network access. Many of the patterns discernible in the early stages of the quantum computing market bore resemblances to those early days of the Internet, or of the Web, mobile computing, open source, big data, and the like. Nothing matched exactly, it never does, but while history never repeats itself it certainly rhymes, and picking out the meter and guessing what comes next in a new emerging tech can be helpful even when one doesn't have deep specialized knowledge. It has helped as I built my understanding of how the quantum phenomena are being shaped into system architectures, as the value chain is beginning to form, as the complete quantum computing stack begins to take shape. There is a vast amount that I do not know and do not understand about quantum computing, but I feel confident that my guesses are not horribly wrong.

Gathering this knowledge, I have found my fascination with quantum physics and this particular moment when its science is being turned into technology has only grown, never diminished. I love the history of physics, the yielding of classical theory to the counterintuitive and often upsetting quantum theory. I love the history of information theory and the dawning of the digital age. My understanding of both has grown and cross-pollinated, mirroring the field of quantum information technology itself. At some point, I was inspired to write this book — to share what I'd found with anyone else who might be drawn in by the same sense of wonder, even without a physics degree.

This book is not a textbook or a how-to of any kind. It will not teach the reader how to program quantum computers. It contains virtually no math, and no code. My hope is that the reader will have a better understanding of what quantum computing is, what this moment in its development is about. This stage of quantum computing's development is, in some ways, mimicking the path taken by digital computing in the middle of the twentieth century, or at least it's rhyming with it. The first functioning devices are being created in a bespoke fashion, with each one an idiosyncratic attempt to find a path towards the usefulness and scalability of the modern microprocessor.

In the midst of this uncertainty and enormous complexity, there can be general confusion and even, sometimes, attempts to leverage confusion as a tool in itself, to gain investment, or revenue, or reputation. This book is an attempt to provide explanatory knowledge to the reader, to help interpret marketing and scientific claims when you encounter them. Mostly I hope your sense of awe and wonder grow a little as you gain a greater appreciation for just how audacious are our goals for this new quantum era.

Who this book is for

I wrote this book to help myself, as a non-physicist, to build a conceptual and contextual understanding of the emerging field of quantum information technologies in computation and communication. My hope is that reading about my own intellectual journey will be of some use to anyone who is curious about how it came to be that quantum mechanics replaced classical physics' understanding of the universe, how it then influenced information theory, and is now transforming the world of digital computing and communication that defines so much of our modern world. It's a complex topic, but I rely on historical, conceptual, and intuitive understanding – there's no math, or virtually no math, and I promise the reader I was completely lost when I first encountered the topic and, over the past decade, I have gotten comfortable with what I cannot understand. I hope you will as well.

What this book covers

Chapter 1, Quantum Quinceañera: Google announces quantum supremacy, and the world says "huh?" IBM reacts with irritation, journalists try to explain with mixed results. What Google was actually saying, what it wasn't, and why the slowdown of Moore's Law makes it matter more than the headlines suggested.

Chapter 2, Don't Watson My Quantum: IBM's long arc from tabulating machines to Watson to quantum is a case study in navigating paradigm shifts. A deep patent portfolio, a disastrous bet on Watson AI, and a surprising early move in quantum — making Qiskit open source — illuminate the tension between proprietary control and open innovation that will define the field.

Chapter 3, A Quantum Mechanical Kludge: You don't have to be a quantum physicist to understand quantum computing, but a conceptual grounding *is* helpful. Classical physics couldn't explain certain phenomena, so some theoreticians made controversial, creative leaps to plug the holes. Planck's "hack," Bohr's atom, the Copenhagen interpretation, Einstein's discomfort, and the eventual vindication of entanglement through Bell's theorem and Aspect's experiments. Despite the controversy, quantum mechanics remains the most successful scientific theory in human history.

Chapter 4, Mid-Century Classical: Physics claims to be the original science, with all other fields derivable from first principles. The mid-twentieth century bolstered the claim considerably. From Babbage and Lovelace through Turing, ENIAC, and von Neumann's open-architecture machine at Princeton, universal computers emerged from a burst of maths, engineering, and a healthy dose of physics — establishing the pattern of codesign between theorists and engineers that will repeat in the quantum era. By 1965, Gordon Moore was observing the rapid innovation that would inspire him to pronounce his "law."

Chapter 5, Gedankenexperiment: IBM's earliest involvement in quantum computing grew from a deceptively simple question: what are the physical limits of computation? We meet some creative and colorful pioneers — Landauer, Bennett, Fredkin, Benioff — and their thought experiments about reversible and quantum computation. This ragtag group convenes at MIT's Endicott House in 1981, and super scientist Richard Feynman pronounces the idea of quantum computers a Good Problem, inspiring a generation to prove him right.

Chapter 6, Universes of Information: Qubits, superposition, Dirac notation, the Bloch sphere, and quantum gates: the mathematical language of quantum computing. Where a classical bit is a switch with two positions, a qubit is a point on a sphere — encoding vastly more information and demanding a richer set of operations. Deutsch's 1985 paper described a theory of universal quantum computing in full, updating the Church-Turing thesis for the quantum age. Whether such a device would actually be *useful* remained an open question.

Chapter 7, Shor, and Everything After: Deutsch and Jozsa published the first quantum algorithm in 1992 — exponentially faster than classical, but solving a problem nobody needed solved. Two years later, Peter Shor at Bell Labs found one that did: the algorithm that breaks all crypto, the one you probably knew about before you picked up this book. Quantum computing was now of interest to more than just some oddball physicists — including the intelligence agencies.

Chapter 8, Experimentalists Just Wanna Have Fun: Shor's algorithm attracted spies and investors, but it did nothing to help build a working machine. A qubit was still a figment of the scientists' imagination. DiVincenzo's criteria bridged theory and experiment, spelling out what a physical qubit must actually do. It took trial and error to find candidates. By 2001, NMR equipment had run Shor's algorithm to factor the number 15 — then NMR was promptly abandoned, the first modality to hit fundamental limits.

Chapter 9, Nature's Perfect Qubits: What really killed NMR was that scientists at NIST had turned an atomic clock into a quantum computer. The story traces from Ramsey and the maser through Dehmelt's and Wineland's Nobel-winning ion trapping work to the first quantum logic gates. Ion trappers like to say every ion is an identical, perfect qubit — nature's own, requiring no fabrication. However, they were slow, and scaling beyond a single chain of ions would prove a formidable challenge.

Chapter 10, Man-made Atoms: In 1962, a 22-year-old postdoc at Cambridge named Brian Josephson predicted that electrons would tunnel across a gap in a superconducting circuit — a prediction that earned him a Nobel Prize. For decades, Josephson junctions were used in scientific instruments. Then, in the late 90s, Nakamura and Tsai at NEC used microwaves to control the state of an artificial atom. They were extremely hard to control and very short-lived, until researchers at Yale developed a design called the transmon that immediately became the dominant superconducting qubit. Meanwhile, D-Wave pursued a controversial alternative path with quantum annealing, to both excitement and skepticism.

Chapter 11, Qambrian Explosion: Making a real quantum computer available to anyone on the Internet in May of 2016 — IBM's Quantum Experience — forever cemented IBM's place in the field's history. But for many, the event served as a starter's pistol. Like the Cambrian Explosion 540 million years ago, quantum computing experienced its own proliferation — competing qubit modalities, startup companies, and national programs erupting seemingly all at once. The private-sector funding surge transformed the landscape, but also raised fears of a quantum bubble.

Chapter 12, The Heroic Era of Qubits: A cadre of physicists, each convinced of their ability to change the world, assembled teams, distilled speculative R&D roadmaps into pitch decks, and convinced sources of capital to invest — all based on uncut self-confidence and little else. These physicist-founders built startups around a technology whose timeline to commercial value was unknowable — a genuine black swan bet. The biggest risk facing the field is not technical failure but a loss of faith in its potential.

Chapter 13, Transcendent Quantum Hype: Tech bubbles are nothing new, but quantum hype has deeper roots than most. The Fundamental Fysiks Group's 1970s fusion of entanglement with counterculture mysticism planted seeds of pseudoscientific association that persist to this day. Combine that with the "simultaneity fallacy" — the persistent misbelief that quantum computers try all solutions at once — and the well-oiled machinery of tech marketing, and quantum computing has been subjected to a hype cycle that threatens to undermine the very progress it celebrates. From D-Wave's "science by press conference" to Michio Kaku on 60 Minutes, the anatomy of quantum hype is dissected — along with the irony that the hippie physicists who fueled the woo also helped prove entanglement was real.

Chapter 14, The Road to Fault Tolerance: What will it take to overcome the noise? The story begins with Von Neumann's engineers testing a thousand surplus vacuum tubes and building for the worst, and with Richard Hamming at Bell Labs devising the first error-correcting codes. But quantum error correction faces challenges with no classical precedent: qubits suffer both bit flips and phase flips, measurement collapses the state, and the no-cloning theorem forbids copying. Shor answered his own critics by inventing the first quantum error-correcting code; Steane found a parallel solution at Oxford; Kitaev's topological codes opened a new frontier. The arc from those theoretical foundations leads through surface codes, bosonic "cat" qubits, and quantum LDPC to the landmark results of 2024-2025 — and the hard questions that remain about scaling from dozens to thousands of logical qubits.

Chapter 15, The Quantum Horizon: Where are we, and where might this be going? The 2025 Nobel Prize validated the journey from Josephson junctions to superconducting qubits. The quantum stock surge — D-Wave up 3,700%, Rigetti up 5,700% — carries echoes of every bubble the book has traced. NISQ machines are proving their worth not as computers but as exotic lab instruments. Quantum sensors are already delivering value. The real applications will emerge from quantum "primitives" we are only beginning to understand, not from the fantasies of the hype merchants. The chapter closes with a return to Deutsch's explanatory knowledge: we are building a new tool for understanding the universe, and the history of such tools suggests the applications will be nothing we can currently imagine.

To get the most out of this book

Relax, stay curious, sit with wonder.

Get in touch

Feedback from our readers is always welcome.

General feedback: If you have questions about any aspect of this book, email us at `customercare@packtpub.com` and mention the book title in the subject of your message.

Errata: Although we have taken every care to ensure the accuracy of our content, mistakes do happen. If you have found a mistake in this book, we would be grateful if you would report this to us. Please visit `www.packtpub.com/support/errata` and fill in the form.

Piracy: If you come across any illegal copies of our works in any form on the internet, we would be grateful if you would provide us with the location address or website name. Please contact us at `copyright@packt.com` with a link to the material.

If you are interested in becoming an author: If there is a topic that you have expertise in and you are interested in either writing or contributing to a book, please visit `authors.packtpub.com`.

Share your thoughts

Once you've read *The New Quantum Era*, we'd love to hear your thoughts! Scan the QR code below to go straight to the Amazon review page for this book and share your feedback.

`https://packt.link/r/1807787370`

Your review is important to us and the tech community and will help us make sure we're delivering excellent quality content.

Free benefits with your book

This book comes with free benefits to support your learning. Activate them now for instant access (see the "*How to Unlock*" section for instructions).

Here's a quick overview of what you can instantly unlock with your purchase:

DRM-Free PDF Version

Download DRM-free PDF and ePub copies of this book.

7-Day Packt Library Access

Get 7-day unlimited access to 8,000+ books and videos. No credit card required.

Available for first-time Packt+ trial users only.

Next-Gen Reader Access

Read this book on Packt Reader with progress sync, dark mode and note-taking.

How to Unlock

Scan the QR code (or go to `packtpub.com/unlock`). Search for this book by name, confirm the edition, and then follow the steps on the page.

UNLOCK NOW

Note: Keep your invoice handy. Purchases made directly from Packt don't require one

Part 1

Theory

Just what, exactly, did Google's "quantum supremacy experiment," announced in 2019, mean? The answer takes us to the early years of the 20th century, when a new framework for understanding our physical world was about to disrupt the field of physics, through the origins of the information technology industry in the middle part of last century, through to the collision of both those worlds in the 1980s, when quantum information theory was born. After the idea was deemed a "good problem to solve" by Richard Feynman, serious effort was applied to creating a complete theoretical foundation that would add quantum mechanics to information theory, paving the way towards quantum computers.

- *Chapter 1 - Quantum Quinceañera*
- *Chapter 2 - Don't Watson My Quantum*
- *Chapter 3 - A Quantum Mechanical Kludge*
- *Chapter 4 - Mid-Century Classical*
- *Chapter 5 - Gedankenexperiment*
- *Chapter 6 - Universes of Information*

1

Quantum Quinceañera

> *It's easy for quantum computing to sound like a fanciful theorist's dream...But just extrapolating Moore's Law...somewhere around the year 2020 or 2030 is when we hit one bit per atom...So if we really want to keep getting faster...Quantum weirdness is essentially the only resource we have that's still untapped for computing.*
>
> — *Isaac Chuang and Neil Gershenfeld*

Google declares quantum supremacy, to general confusion

Google's blog post, "What our quantum computing milestone means,"[1] was written by their CEO, Sundar Pichai, so it was clearly meant to be a big deal. It was quantum computing's coming-of-age moment, the field stepping out of the laboratory and onto the world stage. Conveniently, there's a word for a coming-of-age celebration that starts with "q" — and this is a field with a deep fondness for that letter. That week in October, 2019, it was accompanied by several articles and an op-ed in the *New York Times*, an article and a peer reviewed paper in the prestigious journal, *Nature*, and stories in several other tech outlets. Google had gone to great lengths to generate attention for their accomplishment. It was not, however, immediately apparent just what those claims meant. Pichai called it the "hello world" for the field of quantum computing, something Google had been working at for thirteen years. They had now accomplished this feat on their device named Sycamore, seen in *Figure 1.1*. But to the general public, to the tech journalists covering it, to the vast majority of people, it was far from clear just what Google had accomplished. Pichai wrote in the blog post that his researchers had achieved "quantum supremacy," and that meant that they had "used a quantum computer to solve a problem that would take a classical computer an impractically long amount of time."[2]

This, the reader learned, was a "...milestone in our effort to harness the principles of quantum mechanics to solve computational problems."

Journalists struggled to get to grips with the implications of what was being announced. First things first - what was meant by a "classical" computer? The term is borrowed from physics, where the behaviors at the subatomic scale are referred to as quantum, and above that scale physical behaviors are considered "classical." So, your laptop, your phone, even the largest supercomputer in the world — all computing classically. Which begged the question, what did it mean to build computers using the "principles of quantum mechanics?" In the blog, Pichai had offered the following as a starting point:

> *A bit in a classical computer can store information as a 0 or 1. A quantum bit—or qubit—can be both 0 and 1 at the same time, a property called superposition. So if you have two quantum bits, there are four possible states that you can put in superposition, and those grow exponentially. With 333 qubits there are 2^*^{333}*, or 1.7x10^*^{100}*—a Googol—computational states you can put in superposition, allowing a quantum computer to simultaneously explore a rich space of many possible solutions to a problem.*

A "1" followed by 100 zeros is called a googol, and is the source of the company's name. Including this Easter egg in the form of a very large number, as cute as it was, was not particularly helpful for building understanding. A common characteristic of attempts to explain quantum computing seems to be to utilize analogies that seem straight-forward but don't end up being very useful for building understanding. For example, Pichai's statement that quantum computers can "simultaneously explore...many possible solutions" is not entirely accurate, and leads to gross misunderstanding. And yet these false analogies persist to this day, as even well-informed and competent professionals such as Pichai struggle with the inadequacy of our language to explain what is *actually* going on.

Cade Metz, a journalist writing in the New York Times the same day, side-stepped these challenges, writing merely that the technology "...relies on the mind-bending ways some objects act at the subatomic level or when exposed to extreme cold."[3] Metz' colleague at The New York Times, David Yaffe-Bellany, also weighed in on the same day with *Quantum Computing Explained (in Mere Minutes)*[4] where he grappled with the topic slightly more deeply. Whereas classical computers represented information as "bits," which can only assume the values of "1" or "0" and support computation based on that Boolean representation of two states, true and false, quantum computers used "qubits." Qubits, a portmanteau of "quantum" and "bits,", Yaffe-Bellany explained, can be 1 *and* 0 simultaneously, and therefore could hold an exponentially larger amount of information when compared to an equal number of classical bits.

The definition of qubit is accurate, however, Yaffe-Bellany's explanation of superposition is, like Pichai's, another great example of an attempt to explain a quantum mechanical concept that drives physicists crazy - its inaccuracies leave an entirely incorrect impression. Quantum physicists will tell you that superposition doesn't mean that the qubit has both values at the same time, but instead have a probabilistic combination of both, which doesn't exactly help shed light on the matter. There has been a struggle, largely unsuccessful, to come to grips with the "spooky" behaviors at the quantum scale, and the press coverage of the Google event didn't fare any better. In the *IEEE Journal* qubits were likened to "marbles" instead of coins, with its hemispheres labeled "zero" and "one," rolling around while bits are coins that lie flat with only one value visible[5]. A metaphor that physicists find particularly galling is that quantum computers evaluate "all possible answers at once" before settling on the correct answer, something they decidedly do not do. However, coverage of quantum computing continues to fall back on these flawed explanations.

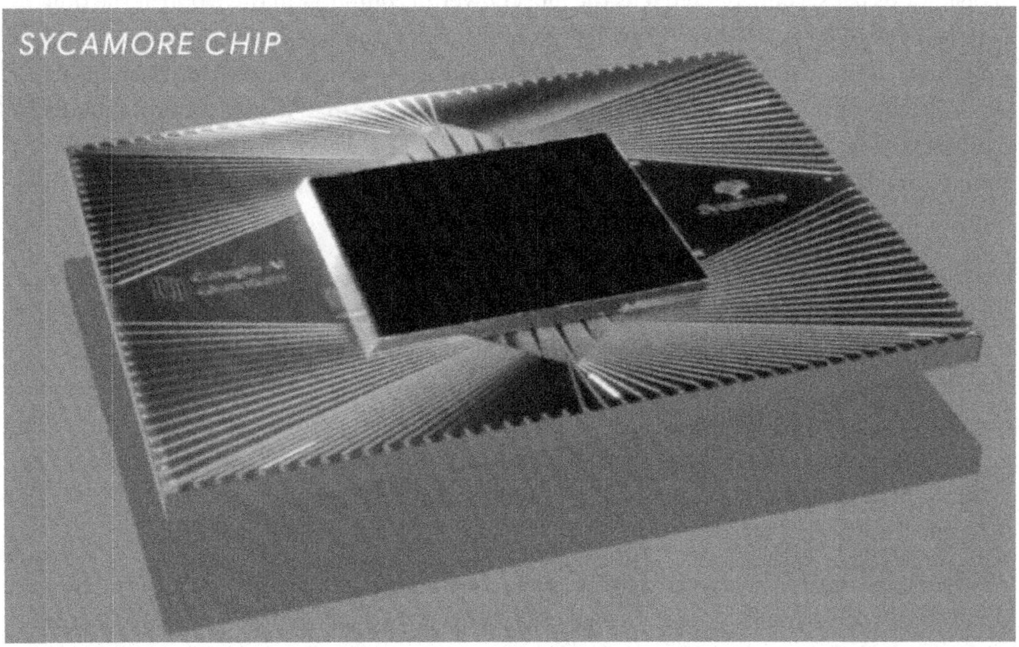

Figure 1.1: The Google Sycamore quantum processor used in the supremacy experiment. By Google, CC BY 3.0, https://commons.wikimedia.org/w/index.php?curid=109155716

The coverage around the Google supremacy experiment, having failed to create metaphors that created more clarity, turned to a framing that was more comfortable — industry competition. Yaffe-Bellany's piece noted that "many of the tech industry's biggest names, including Microsoft, Intel and IBM as well as Google, are jockeying for a position in quantum computing." In fact, the reactions to the supremacy announcement had also featured some uncharacteristically aggressive competitive posturing from tweedy IBM Research in anticipation of the puffy vested Google team's big announcement.

The announcement had been foreshadowed by some drama fueled by that competition. Someone at NASA, a collaborator on Google's experiment, had accidentally posted the research paper to the NASA Technical Reports server a week before Pichai's blog post, where it wouldn't have been noticed if Google's own search engine hadn't ingested it, added it to Google Scholar, where it was fed into the email alerts of anyone monitoring the topic of quantum computing![6] IBM used the inadvertent leak to formulate a rebuttal, seeking to undermine Google's claims, including members of IBM's quantum computing efforts.

IBM had a secret weapon in their effort to undermine Google's claims, a supercomputer at the U.S. Department of Energy's Oak Ridge National Laboratory named Summit. IBM designed and built the machine and had installed Summit in 2018, when its 150 petaFLOPS, or quadrillions of operations per second, ranked it as the most powerful computer in the world at the time. Leveraging their comprehensive knowledge of Summit's architecture, the IBM team hypothesized that, with the help of petabytes of hard drive storage as working memory, it would be able to complete the same experiment in two-and-a-half days instead of hundreds of thousands of years. The rebuttal, posted to IBM's own blog a week ahead of Google's announcement, didn't do much other than make it even harder to make sense of the whole thing.

Regarding investment in the private sector, Yaffe-Bellany cited a Nature article[7] which found that between 2012 and 2019, "venture capitalists have invested more than $450 million into start-ups exploring the technology." On the global stage, quantum computing was characterized as a strategic race at the superpower scale, with "China ... spending $400 million on a national quantum lab," and, "the Trump administration ... promising to spend $1.2 billion on quantum research, including computers." The U.S. investment sounded like an attempt to catch up, because China had "filed almost twice as many quantum patents as the United States in recent years."

This familiar context of a race to be first to market with a new technology, along with the sheer scale of global investment, may have played a role in creating a sense that quantum computing was imminent. The idea that this was the "hello world" moment for quantum computing, a phrase used by Pichai and repeated by the article[8] in *Nature,* certainly helped create this impression. The true nature of the experiment seemed as difficult to explain as the quantum behaviors that underpinned the workings of the machine. In fact, the experiment was designed to take advantage of that "quantumness" in a way that would be particularly challenging for a non-quantum classical computer to emulate.

The most cogent explanation of just what Google's announcement meant came from mathematician Scott Aaronson, the director of the Quantum Information Center at the University of Texas in Austin. Aaronson had been involved in the design of the experiment carried out by Google's research team, and had a reputation as a science communicator and prolific blogger. Writing an op-ed in the New York Times a week after the initial announcement, Aaronson set the context for understanding appropriately, writing "the calculation doesn't need to be useful: much like the Wright Flyer in 1903, or Enrico Fermi's nuclear chain reaction in 1942, it only needs to prove a point."[9] Drawing parallels to both the invention of flight and the atomic bomb ultimately seemed to further confuse the search for clarity in the press.

The challenges faced by the journalists covering the event and their readers contributed to a fundamental misapprehension about the "supremacy" moment — that the technology was on the cusp of emergence, and that in some reasonable amount of time it could be expected to be useful for something, and further, that it was an evolutionary step forward for the information technologies with which we were already familiar. This is the understanding of how information technology works since the late twentieth century — a new chip, a new application, faster wireless networking, better functionality, some other measure of progress through innovation makes its way into the media, into our consciousness, and a year or three later, there it is in the store or on the e-commerce site, or in the App Store. Generations of technology iteratively improve on their predecessors and Apple Newtons are supplanted by Palm Pilots, then RIMM Blackberries which are in turn kicked to the curb by the iPhone.

It's a comfortable narrative because it's familiar, but that doesn't make it appropriate framing for understanding what quantum computing was in 2019. It's not like there wasn't sufficient backstory for the Google accomplishment to signal an arrival, a coming out - *Time Magazine* had run a cover story in 2014, five years earlier, heralding the launch of D-Wave's "Infinity Machine,"[10] the first commercially available quantum computer. The company, based in British Columbia, Canada, claimed they would soon be capable of cracking all computer encryption schemes. It was true, the math that made this feat theoretically possible did exist — we will explore it in *Chapter 7* — however, given the limitations of all known quantum devices, the claims in the article were seen as highly dubious by the quantum physics and quantum computing communities. Even at press time, D-Wave's claims of operating on the quantum level had been under fire by scientists, and by the time of the Google announcement in 2019 no one had managed to produce anything like their proof of quantum performance using D-Wave's machine. However, given the time elapsed between the *Time* story on D-Wave and this Google accomplishment, it seemed reasonable to believe the technology had matured considerably.

But no matter what impression the collective coverage of the experiment conveyed, the reality of quantum computing was far less familiar, and far less mature. What the Google team, and the IBM team, and similar teams at various tech giants and startups, and researchers in academic labs around the world were working towards was actually a complete paradigm shift, in literal terms. The machines they were building were attempting to rewrite the foundational information theory that ruled every transistor on every chip, every instruction in every piece of software, every interaction with a computer since computers were invented in the middle of the twentieth century. As Aaronson wrote, quantum computing was an attempt to rewrite the "rules that Charles Babbage understood in the 1830s and that Alan Turing codified in the 1930s," rules adhered to by "every computer on the planet — from a 1960s mainframe to your iPhone."[11] If successful, their machines would create solutions to classes of problems that were deemed impossible to solve using the computers no matter how powerful we made them. IBM's Dario Gil had his own take on Pichai's playful "googol" explanation, in an interview with the NY Times that was packaged with IBM's rebuttal. He said that if one were to build a system with 100 perfect qubits, "you would need to devote every atom of planet Earth to store bits to describe that state of the quantum computer."[12]

In 2019, however, no-one even had built a single perfect qubit, nor does one exist at the time of this writing. The Google experiment was contrived, pointless from the perspective of the kind of useful things we expect from computers. Tasked with anything we'd recognize as useful; the best quantum computer would lose in a head-to-head battle with a 10-year-old laptop. It wouldn't even be close.

So why the excitement? Why all the investment? The context the media missed or glossed over was that the incredible progress delivered by classical computing has, in the past decade, been showing signs of slowly or even coming to an end, which struck fear into the captains of tech and the heads of state alike. A growing sense of panic was gripping the tech world, causing it to cast about for a new engine of growth. Despite quantum computing being nowhere near ready for its mainstream debut, the industry was looking farther and farther afield for a solution to a looming crisis. Moore's law, which described the massively powerful driving force behind decades of technological advancement and economic production, was showing signs of falling apart. Computer industry insiders were increasingly concerned about finding ways to maintain their progress, and quantum was emerging as a potential savior.

In fact, Moore's law is not a law at all. In 1965, Gordon Moore, the director of research and development at Fairchild Semiconductor, wrote an article in which he observed that the number of components on a chip had doubled each year since 1958, leading to roughly doubling what a chip could do as well. He expected this trend to continue for at least another 10 years. A decade passed, and Moore reassessed his prediction at a conference, where he said he believed the trend would continue, revising the period of doubling to every two years.

Carver Mead, a famed professor at Caltech, coined the term "Moore's law," and it stuck. It's a fixture in the language, largely because it was demonstrated by the entire chip industry for the next 30 years, as illustrated in *Figure 1.2*.

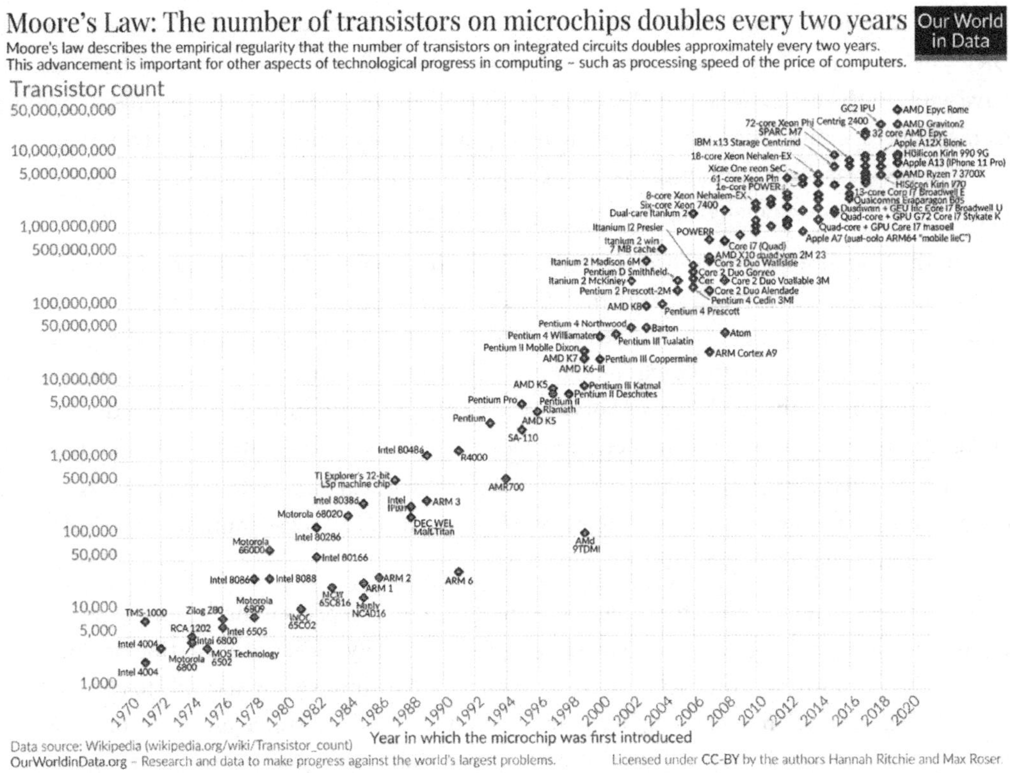

Figure 1.2: A logarithmic graph showing the timeline of how transistor counts in microchips are almost doubling every two years from 1970 to 2020; Moore's Law. Max Roser, Hannah Ritchie - https://ourworldindata.org/uploads/2020/11/Transistor-Count-ov'ser-time.png

Moore's law describes the progression from the Intel 8088 chip in 1980 with 29,000 transistors to Intel's Itanium chip in 2000 having 220 *million* transistors. In many ways global trade, economic growth, and geopolitics have all been linked to this reliable, remarkable progress. The progress has been so remarkable that it seems to defy probability, with predictions of its failure having been made again and again, only for a new chip to deliver yet another leap forward. However, in the early 2000s the concerns began to accumulate. Sluggish economic growth was attributed by some to the slowdown of microprocessor advancements.[13] The stakes extend well beyond the economy. The great challenges facing humanity — addressing the climate crisis, discovering new materials, understanding the molecular machinery of life — all demand computational power that may be approaching the limits of what classical technology can deliver. If our toolmaking stalls, so does our ability to sustain ourselves.

As these concerns grew, the search for alternative technological approaches has gotten more attention. Exotic ideas like designing chips to operate more like our brains[14], computing using DNA[15] or even fungi[16] have been able to mobilize more funding for research than they may have without the concerns about Moore's Law. Quantum computing is a very compelling idea from this perspective, as its mathematical ideas provide a neat solution to a fundamental bottleneck of classical computing that is present in certain types of problems that get exponentially more challenging as they increase in scale.

In fact, the Google experiment was explicitly designed to demonstrate this core characteristic of quantum computing. Its chip, known as Sycamore, had 53 operational qubits, on which a random set of quantum operations was executed, the equivalent of a very small computer program, and then their output was read out. A collection of quantum operations is known as a "circuit" to those in the field. Because of the way Google's Sycamore chip exploited the characteristics of quantum mechanics, representing all the possible states of the circuit, and therefore calculating the same output, would take 2^{53} classical bits, or 9 *quadrillion* bits. By comparison, the most advanced microprocessors as of the mid-2020s contain on the order of 100 billion transistors. A quadrillion is a million billions.

This was the reasonable basis for Google's claim that they had proven the "supremacy" of quantum computers over classical. Notwithstanding IBM's hypothesized solution on Summit's 250 petabytes of storage, the fact remained that it took Sycamore a few seconds to complete the task, still several orders of magnitude faster than the classical supercomputer. In the years that followed, classical researchers would continue to chip away at Google's claimed gap, finding cleverer ways to simulate the original Sycamore experiment. This back-and-forth would become a defining dynamic of the field — each quantum advance met by classical counterpunches, raising the bar for what "advantage" truly means.

The Google experiment established a credible claim that supremacy would eventually be provably real, no matter how regrettable the use of that particular word. It's worth noting that none of the coverage of Google's experiment seemed to take notice over the use of a word with prejudicial associations. Since then, many people have expressed their dislike of the word "supremacy" to describe the milestone, and John Preskill, the term's originator, has voiced his own regrets on numerous occasions. A preferred term has emerged, "quantum advantage," but since Google used the supremacy term originally, it's less confusing to stick to "supremacy" when discussing the experiment. By 2025, "quantum advantage" had become nearly universal in the field, and Google themselves adopted the phrase "beyond classical" when announcing their next-generation Willow chip in late 2024. The language is still evolving, as are the particular metrics by which this advantage will be judged. The underlying ambition remains the same: to build a quantum computer that does something useful that a classical computer cannot.

This oversight regarding word-choice aside, the news and analysis around the experiment left much to be desired. Over-reliance on frames of reference from decades of emerging technologies, tech rivalries, venture capital investment trends, and global economic dynamics caused most pieces to miss the real news in the announcement. Appreciating the breakthrough requires a deeper dive into exactly what quantum computing is, what makes it so radically different from classical computing, its relationship to the origins of quantum physics, and how adding quantum mechanics to information theory is opening the door to an entirely new way of computing.

Quantum computation may provide a way to sustain the trajectory of Moore's law, not by continuing the relentless doubling of transistors but by replacing classical bits with quantum ones. It is hoped the quantum gambit will allow us to side step the plateau with an entirely different technology that may break classical computation bottlenecks and open up miraculous new applications in bioscience, material science, communications, and financial services, among others. These machines' ability to simulate natural systems with far greater fidelity than any classical system was hoped to lead from the demonstration of a contrived "supremacy" ultimately to remarkable breakthroughs in our understanding of our universe itself. But first, the field was going to have to move past their Kitty Hawk moment and prove there was a path to a productive, useful technology. This was going to require an unprecedented effort to bridge the worlds of theoretical physics with engineering and technology development, and effort that had begun decades before Google's results in 2019.

References

1. 'What Our Quantum Computing Milestone Means --- Blog.Google', n.d. https://blog.google/technology/ai/what-our-quantum-computing-milestone-means/
2. Google blog https://blog.google/perspectives/sundar-pichai/what-our-quantum-computing-milestone-means/
3. NY Times https://www.nytimes.com/2019/10/23/technology/quantum-computing-google.html
4. NY Times https://www.nytimes.com/2019/10/23/business/quantum-computing-google.html?searchResultPosition=3
5. IEEE Spectrum https://spectrum.ieee.org/how-googles-quantum-supremacy-plays-into-quantum-computings-long-game
6. IEEE Spectrum What Google's Quantum Supremacy Claim Means for Quantum Computing - IEEE Spectrum
7. https://doi.org/10.1038/d41586-019-02935-4
8. Nature https://www.nature.com/articles/d41586-019-03213-z/
9. NY Times https://www.nytimes.com/2019/10/30/opinion/google-quantum-computer-sycamore.html
10. Time Magazine https://content.time.com/time/subscriber/article/0,33009,2164806,00.html
11. NY Times https://www.nytimes.com/2019/10/30/opinion/google-quantum-computer-sycamore.html
12. NY Times https://www.nytimes.com/2019/10/21/science/quantum-computer-physics-qubits.html
13. MIT Technology Review https://www.technologyreview.com/2020/02/24/905789/were-not-prepared-for-the-end-of-moores-law/
14. MIT Technology Review https://www.technologyreview.com/technology/neuromorphic-chips/
15. MIT Technology Review https://www.technologyreview.com/2021/10/27/1036819/dna-computing-spintromics-history/
16. Royal Society https://royalsocietypublishing.org/doi/10.1098/rsfs.2018.0029

2

Don't Watson My Quantum

Once a new technology rolls over you, if you're not part of the steamroller, you're part of the road.

— Stewart Brand

Competitive dynamics in quantum rooted in AI and earlier, in its origins

IBM's peevish "well, actually" response to Google's supremacy experiment went beyond merely objections to the experiment's design and their resulting claims. The company felt it had been crucially involved in the roots of the field, with active investment that went back at least twice as long as this bunch of Silicon Valley startup kids. Given its legacy of contributions right down to coining the phrase "quantum information science," one could understand how annoying it would be to have to call "dibs" on the emerging technology. There was another emotional response that Google's move to grab the quantum spotlight elicited, a deep gut reaction culturally unique to the IBM of the 21st century, best described as "oh no, not again."

While IBM itself has a history that stretched back over a hundred years as a manufacturer of record-keeping and measuring systems like meat scales and punch card readers founded in 1911, its identity as a computer company began with its IBM Research division. Founded in 1945, contemporaneously with the birth of classical computing, IBM's original research teams focused on the fundamental science that underpins the technologies derived from information theory. As the effort grew from its origins in a lab at Columbia University to purpose-built buildings of its own, the research division leveraged chemistry, physics and other disciplines to create intellectual property and earn Nobel and Turing prizes. In 1961 IBM built the T.J. Watson lab in Yorktown Heights, NY, designed by the mid-century modern architect Eero Saarinen (see *Figure 2.1*), which serves as the Research headquarters to this day. In many ways it is the last of the great industrial R&D powerhouses at General Electric, Westinghouse, Bell Labs, Xerox Parc and others that drove the transition from the late industrial era to the early Information Age. The Research division filed more patents than any other firm every year from 1993 until finally losing the top spot to Samsung in 2022, after an unparalleled run of twenty-nine years.

Figure 2.1: IBM's T.J. Watson Lab, Yorktown Heights, NY

Fueled by Research's intellectual property output, the IBM corporation had defined the early part of the information technology industry with its "mainframe" computers entering the market in 1952. The mainframe product family converged under the System/360 architecture and brand in 1964, and by 1973, IBM's $11 billion in revenue[1] represented one third of the entire U.S. technology market[2]. IBM powered banks and governments and every other large-scale computing application so ubiquitously that "no-one lost their job buying IBM" entered the business lexicon. In parallel, the Department of Justice investigation into alleged anti-competitive monopolistic practices, opened in 1969, gathered steam.[3]

In the 1980s, a new product for a new market presented itself as a way to take the DOJ heat off IBM by demonstrating their largesse to the little fish swimming around in the same market. The product was the personal computer or "PC," a device with a tiny fraction of the power of the machines marketed to industry and governments, for a fraction of the cost. The IBM PC had been developed as a kind of "skunkworks" project, not on the formal R&D roadmap but tolerated as an oddball experiment. Skepticism about the size of the market for the PC created a constant battle for resources to get it made, but by the late 70s it was nearly ready for launch.

A question remained about what the operating system for the PC would be, and here an executive decision was made that would change the course of history not just for IBM but for the entire world. Part of the pressure from the DOJ around the mainframe arose from the fact that IBM controlled both the hardware and the software and no competitor could gain even the tiniesttoehold in the market without risking being sidelined by IBM. The PC was already leveraging off-the-shelf hardware components built to specifications to create a more open system, and so why not license the operating system as well? Enter Microsoft, and exit the IBM technological hegemony, and with it a vital piece of IBM's own perception of its strength and capabilities.

Though it would take decades, the mainframe and the very notion of massively centralized computing power would slip from dominance to a footnote in the IT industry, and IBM slowly slipped down the power rankings. The dominance of distributed computing across virtual servers in the cloud since the 2000s has largely been the story of Google and Amazon Web Services, with Microsoft as a fast follower. IBM's cloud offering is not seen as defining the market in any way, which is particularly galling for IBM. The concept of a virtual machine, crucial to the modern cloud, was seen to have emerged in the late 1990s by Internet service providers hosting multiple customers' web sites on one server. IBM had, in fact, shipped code to run the first virtual machine on a mainframe host in 1967, but had failed to capture any of the value created by that innovation.

As a wellspring of intellectual property, IBM Research was seen by many in the company as the likely source of innovations that could recapture market leadership. However, the one model of generating commercial value from intellectual property that IBM Research knew was outbound licensing, the practice of allowing others to use patented knowledge to create commercial products for a fee. In the early 2000s IBM's income from IP licensing peaked at nearly $2B per year, a powerful stimulant that ensured the mountain of patents would continue to grow. Internally, however, commercializing its own IP proved an elusive goal. Research teams were tasked to work more closely with product teams, in the hopes they could create novel IP just over the horizon on existing product roadmaps, with little success. Experiments with various incubation and acceleration techniques were similarly fruitless. Its two greatest market triumphs in the late 90s and early 2000s — the WebSphere application server and its Power line of servers — owed their market success to others' innovations, Sun's Java programming language and the open source Linux operating system, respectively.

Hope for IBM Research to deliver a win accumulated around artificial intelligence projects. Advances delivered by the division in the mid-1980s in the use of advanced statistical analysis had been successfully commercialized by the company in products like dictation software and automated voice response phone systems, and IBM was seen as a market leader in that niche. The statistical and probabilistic approaches to artificial intelligence continued to evolve, resulting in IBM's Deep Blue software defeating world champion player Gary Kasparov in a chess match in 1997.[4] Proficiency at chess has been an benchmark for artificial intelligence systems since their very first inception in the 1950s, so this was an historic achievement.

Building on what was learned from Deep Blue, IBM Research then focused on more general models for simulating intelligence, and delivered a ground shaking Jeopardy win in 2011[5] with a system IBM named Watson, after the company's founder, Thomas J. Watson, who also lent his name to IBM Research's headquarters. Under CEO Ginni Rometti's leadership, IBM Watson became a "bet the farm"[6] play in the 2010s, with enormous marketing campaigns and the creation of an entirely new division of the company underscoring how big a deal this was seen to be.

There was certainly reason to be enthusiastic. Artificial intelligence had been the dream from the dawn of the industry when early devices were often referred to as "electric brains." With Deep Blue and Watson's triumphs over human minds in chess and general trivia, it seemed that the time was right to deliver significant commercial value from the technology. For a number of years it seemed that IBM might have a winning horse in the race.

Then Google entered the chat. A London-based company named DeepMind, acquired by Google in 2014, formed the nucleus of a major effort in AI in a division called Google Brain. In 2016, DeepMind's AlphaGo software beat the world Go master at the game, a feat many, many times more challenging than beating humans at chess.[7] AlphaGo used a very different approach than Watson, based on a "neural network," a software approximation of the way we believe our own brains learn to recognize and act on patterns. Neural networks weren't a new idea by any means, rooted in our first information theory models for understanding human reasoning as far back as the late nineteenth century. The idea of an artificial neuron originated in a 1943 paper by Warren McCulloch and Walter Pitts[8], and various attempts to build more sophisticated versions paralleled the development of computers from the very beginning. However, in many ways the theory had outstripped the available computational power by the mid-1980s, and attention had moved on to other approaches. What DeepMind had realized was that they could now marshall enough computational power to train a neural net powerful enough to tackle the challenge of Go. Played on a 19x19 grid, a game of Go has 10^{360} possible moves, an unimaginably large number which dwarfs the 10^{123} possible on chess' 8x8 board. Using approximately 50 NVIDIA's GPUs originally designed for computer graphics, repurposed for the math that drives neural networks, AlphaGo was trained over a period of weeks on a diet of millions of Go matches to recognize patterns that would suggest a probable path to a win.

The triumph of AlphaGo over the world champion human player by itself did little to upset IBM's dominant position in the emerging AI market. Buzz in the mainstream press isn't what sells enterprise software, after all – perceived value and inherent trust are required to compel a CIO to sign on the dotted line of a software contract. And IBM felt very confident about the market perception of their trustworthiness, as the enormous uncertainty about this new technology brought the "no one lost their job buying IBM" lore back to the forefront.

However, in the era of the Internet and open source, something had happened to the decision-making power dynamics of the IT market, something that IBM simply didn't have the corporate DNA to understand. Instead of a strictly top-down decision-making process driven by the C-suite, technology investments had become increasingly driven by software developers themselves. Faced with a challenge at work, devs would tend to find potential solutions in technologies downloadable from the Internet, try them out informally and develop solutions that were deployed without centralized approval or even knowledge. The "world wide web" itself had spread that way, with the first web server software packages like *httpd*, which was made available as a source code download from the University of Illinois in 1993[9].

Grassroots technology adoption was embedded in Google's corporate DNA, its origin story in a dorm room at Stanford University was a canonical example of the phenomenon[10]. And so, it was natural for Google, on the heels of the AlphaGo win, to release free, open source tools for developers to build their own neural nets. Called Tensorflow, the software developer kit (SDK) was released in 2015 and almost instantly began to take the air out of IBM's Watson balloon, reaching a stunning 89% market share of AI projects by 2018[11]. Suddenly IBM was in catch up mode, again.

All this was the backdrop to IBM's reaction to Google's 2019 quantum computing supremacy announcement. The supremacy experiment seemed like AlphaGo toppling Watson all over again, a demonstration of capability in an area where IBM took pride in its first-to-market technology that might be followed by a giant sucking sound as the developer community flocked to Google's platform and left IBM high and dry.

There was reason to feel territorial about the fledgling quantum computing market as well — not only had IBM launched its cloud service – the amusingly named the IBM Quantum Experience – in 2016, along with its commercial offering, and an open source SDK, Qiskit, its contributions reached all the way back to the very beginning of the field. In fact, two years before Google's announcement of the supremacy experiment, IBM Research had staged its own quantum computing triumph at an event they called the Think Summit, held in December 2017.

At the event, the pioneers of the field presented their work on algorithms and theoretical proofs of the performance advantages quantum computers would bring over classical technologies. Its primary function, however, was to build on the progress IBM had made in recent years, building on the offering on the public cloud, the SDK and the commercial offering by announcing they were working on a system with 53 qubits, several times larger in scale than anything that anyone had built to that point. The system publicly launched in October of 2019[12], the same month Google announced its supremacy experiment. It came like a bolt out of the blue, taking the IBM quantum team completely by surprise. IBM felt it had a lock on the technology, after all, it had been involved with the very earliest stages of developing the notion that quantum phenomena might provide some advantage for computation.

In fact, the Think Summit was consciously an echo of an event IBM had helped organize decades earlier, in 1981, a conference known as "The Physics of Computation."[13] The event brought together a small group of people who had individually been poking around at the edges of an idea that quantum physics may be useful to incorporate into information theory. The idea had been emerging over the previous decade, and this conference had the explicit goal of determining whether it had any merit. The gathering is universally recognized as the launch pad for the field of quantum computing. It was this history that gave IBM its sense of clear leadership in the field, leadership that Google now was threatening just as it had in AI. The fear experienced around the supremacy announcement was that the firm's efforts in quantum would suffer the same fate as Watson.

Competitive threats were far from the only risk facing anyone vying for the quantum computing crown. The technology was not merely nascent, it was still mostly theoretical, with dependencies on many science breakthroughs still to come. At the final session of the 2017 IBM summit, Dario Gil, director of IBM Research and champion of IBM's investment in quantum computing, moderated a panel of experts. A pioneer in the field, UK-based physicist Richard Jozsa co-creator of the first quantum algorithm in the 90s, was present. Gil asked the panel what they would like to see done with this new marvel, the 53-qubit system to be launched in the coming year. Jozsa said he'd like to see the fundamental rules of quantum mechanics tested on the machine, because he wasn't convinced we had the physics right.

References

1. https://www.ibm.com/ibm/history/history/year_1973.html
2. Richard S. Bower, "Market Changes in the Computer Services Industry," Bell Journal of Economics and Management Science 4 (Autumn 1973): 541-42.
3. https://www.justice.gov/atr/case-document/united-states-memorandum-1969-case
4. Bruce Weber, "Swift and Slashing, Computer Topples Kasparov," The New York Times, May 12, 1997, https://www.nytimes.com/1997/05/12/nyregion/swift-and-slashing-computer-topples-kasparov.html.
5. John Markoff, "Computer Wins on 'Jeopardy!': Trivial, It's Not," The New York Times, February 11, 2011, https://www.nytimes.com/2011/02/17/science/17jeopardy-watson.html.
6. Adi Ignatius, "Don't Try to Protect the Past," Harvard Business Review, July 17, 2017, https://hbr.org/2017/07/dont-try-to-protect-the-past.
7. https://www.scientificamerican.com/article/how-the-computer-beat-the-go-master/
8. https://link.springer.com/article/10.1007/bf02478259
9. https://en.wikipedia.org/wiki/NCSA_HTTPd?wprov=sfti1
10. https://web.archive.org/web/19971210065425/http://backrub.stanford.edu/backrub.html
11. https://d1.awsstatic.com/whitepapers/nucleus-tensorflow-2018.pdf
12. Martin Giles, "IBM's new 53 qubit quantum computer is the most powerful machine you can use," Technology Review, September 18, 2019, https://www.technologyreview.com/2019/09/18/132956/ibms-new-53-qubit-quantum-computer-is-the-most-powerful-machine-you-can-use.
13. 'Celebrating the 40-Year Anniversary of the Physics of Computation Conference | IBM Research Blog --- Research.ibm.com', n.d. https://research.ibm.com/blog/qc40-physics-computation.

Get this book's PDF version and more

Scan the QR code (or go to `packtpub.com/unlock`). Search for this book by name, confirm the edition, and then follow the steps on the page.

UNLOCK NOW

Note: Keep your invoice handy. Purchases made directly from Packt don't require an invoice.

3

A Quantum Mechanical Kludge

"Those who are not shocked when they first come across quantum theory cannot possibly have understood it."

— Niels Bohr

The universe turns out to be quite a bit stranger than we'd hoped.

That the IBM event, intended to celebrate the progress towards useful quantum computers, would culminate in a challenge to the axioms of the field itself might seem odd, but the controversy was not new. In fact, from its inception as a collection of answers to the most vexing problems faced by physics at the turn of the twentieth century, quantum mechanics has struck many scientists as plainly impossible and incorrect, or at least incomplete. This may have something to do with the fact that the answers were struck upon out of a desperate attempt to make our explanations of the universe work.

Humans are very odd animals. Not content with securing requisite calories and water for continued existence, at some point in our hidden collective past we decided that a good thing to do with our unusually large brains was to ask the question, "why?" From the accumulated observations of ancient astronomers to the insistence of Thales of Miletus that all natural phenomena must have a natural cause, we slowly built a tradition of explaining nature through experiment and mathematics rather than myth. By the time Galileo and his contemporaries had placed the sun at the center of our solar system — validated by observation, not scripture — physics had established itself as a discipline capable of extraordinary predictive power. The origins of the engraving seen in *Figure 3.1*, used by Flammarion in a book in 1888 are unclear, but it appears to show a medieval natural philosopher piercing the boundary of a heliocentric cosmology to find the various systems of nature beyond the veil of superstition and myth.

Figure 3.1: "Flammarion engraving" c. 1888

Given the impressive progress and scope of the physicists' ability to create logical, mathematical models that precisely captured and predicted the behavior of nature, the growing confidence in the power of the field should come as no surprise. Rationalism as a philosophical belief formally emerged in continental Europe in the 17th century, rooted in the idea that, given a set of foundational principles such as the axioms of geometry, one could work outwards to deductively reproduce all other knowledge. German Enlightenment-era polymath Gottfried Leibniz and Dutch philosopher Baruch Spinoza in particular believed that a reliance on empiricism, or mere measurement of the world, was a necessary crutch for our limited human minds. God was thought of as an invisible watch maker, and the universe his magnificent watch works. In order to understand God, man must create an empirical understanding of His works.

The Enlightenment as a Western European cultural era was the embodiment of rationalist thinking and, at its most optimistic, held a growing optimism in humankind's eventual completion of the project to create a complete and predictive model of the universe that would all just made sense. The generic pattern employed by natural philosophers such as Leibniz and Sir Isaac Newton was to either devise an experiment to demonstrate a behavior of nature, or to work out the mathematical equations that could explain and predict that behavior. This division of tasks persists in the field of physics today, with scientists identifying as either experimentalists or theorists.

Sometimes, however, the experiment and theory wouldn't line up. These discrepancies have long been irresistible bait for natural philosophers of old and modern physicists alike, since they flag areas of our understanding that are incomplete or incorrect, though fixing those inconsistencies can be very challenging. Observations of Mercury's orbit made in the middle of the 19th century weren't theoretically understood until Einstein's general theory of relativity fifty years later. Another example of curious experimental results took even longer to explain, and resulted in perhaps the largest disruption to our scientific understanding of nature.

In 1801 Thomas Young, a British polymath known as "the last man who knew everything"[1] by his contemporaries, sought to improve the understanding of the interference phenomenon, in which light can interact with itself to create puzzling results. The prevailing theory of light as "corpuscular," or made of particles, had been described by Sir Isaac Newton in his book *Opticks,* published in 1704. However, if light particles collided they would bounce off each other like marbles, which could not explain some of the observed behaviors of light, and this contradiction motivated Young to explore further.

Starting with a shallow vessel of water divided in two by a wall with two small openings, he demonstrated how waves passing through those openings will create two sets of ripples fanning out from the other side of the barrier. Unlike marbles, waves colliding didn't simple bounce off each other. Where the waves overlap with, an interference can be observed. If the waves are aligned, they will create a new wave with a higher peak or deeper trough. If unaligned they will diminish one another, even to the point of rendering that point of water perfectly level if they are evenly and oppositely matched. These are called constructive and destructive interference, occurring when the waves are said to be in superposition with one another.

Young then moved on to light, creating an opaque barrier with two slits for light to pass through, just as the waves in water had through the gaps in the tank wall. When observing the light illuminating a surface on the other side of the barrier, Young could see clear lines of interference. The bands of light and dark showed the light passing through each opening had fanned out just as the waves of water had, overlapping with one another and, when superimposed in that way, had either amplified or diminished the intensity of the resulting light. This double slit experiment showed light behaving as waves in water, posing a serious challenge to Newton's corpuscular theory of light.

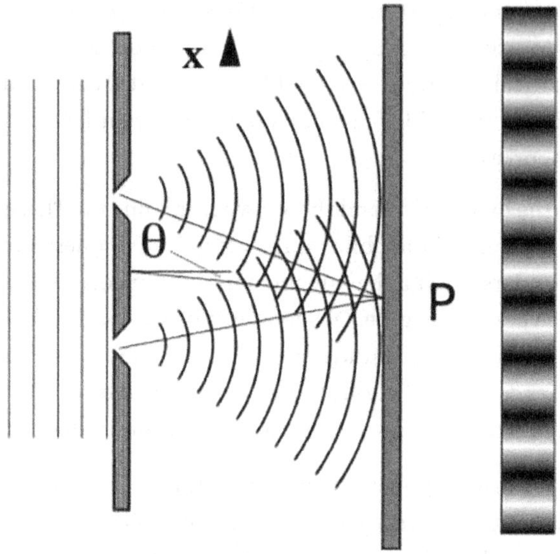

Figure 3.2: The double slit experiment, using two parallel openings in a barrier, a light source, and a flat plane, demonstrates how particles of light, or photons, can behave as waves and interfere with one another as ripples in water would, creating an interference pattern on the plane.

A replacement theory would take another 60 years, until James Clerk Maxwell, a Scottish physicist working on a comprehensive theoretical foundation of electromagnetic behavior, realized that he could explain Young's experimental results and many other behaviors of light using equations developed for electromagnetic phenomena. The publication of Maxwell's *A Dynamical Theory of the Electromagnetic Field* [2] in 1864 is an ideal example of scientific progress in physics, unifying disparate theories and experimental results, producing a foundation large and stable enough for a huge scope of exploration, leading to additional breakthroughs and even paving the way for future technologies such as radio, television, and telecommunications. The Rationalist dream would have been for work such as Maxwell's to perpetuate throughout physics, unifying theory and explaining experimental results, culminating in a complete theoretical model of our universe that can predict any natural system's behavior.

However, the universe had other ideas. Scientific inquiry along a particular thread initially plucked by French civil engineer and physicist, Augustin-Jean Fresnel in the first years of the 19th century, well in advance of Maxwell's work, would eventually lead to serious challenges for the Rationalist way of thinking. Aside from inventing the lens used in lighthouses to increase their range and visibility, Fresnel's contributions to optics included quantification of the interference of Young's double slit experiment and other aspects of the wave-like behaviors of light.[3] Presenting his work to the French Academy of Sciences in 1819 eliminated support for Newton's corpuscles once and for all. Light was now understood to exist as a wave, and as such fit into the broader wave function theory that culminated in Maxwell's classical wave theory. Light was a fully integrated phenomenon in the broader electromagnetic spectrum of radiation. However, not all the details of this comprehensive theory actually made sense.

In 1862, German physicist Gustav Kirchhoff, working on the interactions of electromagnetic radiation and thermal energy, created an idealized, abstracted object he called a "black body," which was non-reflective, opaque, and of uniform density. Its simplified form could be a stand-in for planets or other real-world objects that could be observed to emit heat when exposed to light or other electromagnetic radiation. What Kirchhoff was seeking was a mathematical model that would predict the amount of heat emitted in proportion to the energy of the radiation to which it is exposed. Experimentalists such as Scottish meteorologist and physicist Balfour Stewart picked away at the problem, trying to make enough observations for the theorists to stitch together into a generalized mathematically-provable theory. By the early years of the 20th century, a theory named after British physicists Lord Rayleigh and Sir James Jeans, the Rayleigh-Jeans law, provided an accurate model for black-body radiation, but only for at longer wavelengths. Extrapolating to the shorter wavelength radiation, the calculated energy emitted becomes infinite, an obvious flaw in the theory that became known as the "ultraviolet catastrophe,"[4] illustrated in *Figure 3.3.*

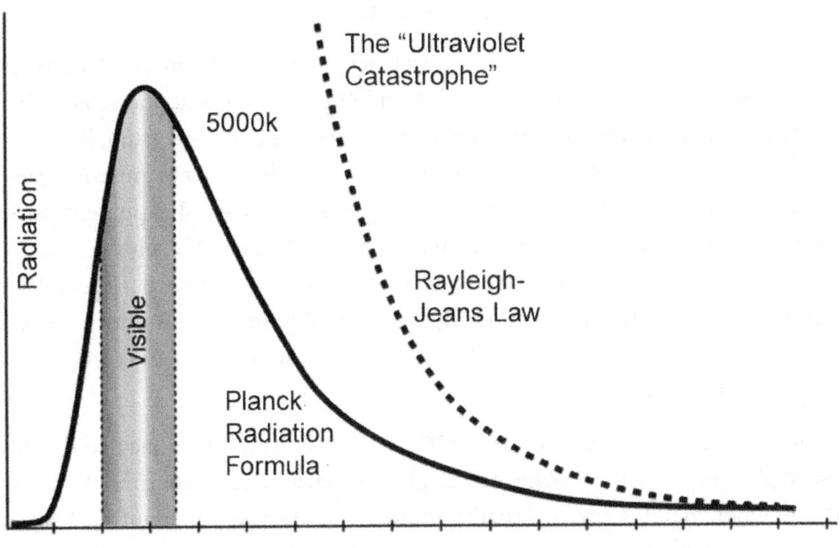

Figure 3.3: A plot of the incorrect calculations provided by Rayleigh-Jeans law which result in radiation energy going infinite, a physical impossibility. http://hyperphysics.phy-astr.gsu.edu/hbase/mod6.html#c4

There was a solution on the horizon for this problem of the Raleigh-Jeans law's incorrect result, however, it would open a door that would fundamentally undermine physics as understood in the 19th century and create knotty problems for the Rationalist view of the universe that remain in many ways unresolved to this day. The originator of this creatively destructive solution was German physicist Max Planck, who would be awarded the Nobel Prize in 1918 for his breakthrough. Leveraging a mathematical framework developed by Austrian physicist Ludwig Boltzmann for his work in statistical mechanics, he posited that energy existed in discrete quantities, rather than a continuous field. This was no small leap — though Planck himself said he didn't think deeply about the theoretical implications as a threat to classical physics, calling his postulate about energy quantization, "an act of despair ... I was ready to sacrifice any of my previous convictions about physics".[5]

Planck's "act of despair" consisted of applying a kind of brute force hack to the theoretical model of black-body radiation to make it work. The recasting of energy as something that comes in discrete "quanta," or quantities, had no foundation in the principles of electromagnetic radiation, and in fact was not done in response to Raleigh-Jeans, as it is sometimes framed in textbooks. It was merely Planck's "hack" to make the model work, not Planck flipping the table on the classical world. In the same letter he went on, "a purely formal assumption ... actually I did not think much about it ..." Quantum mechanics was, in a very real sense, a kludge — a series of improvisations by physicists who needed the math to work and were willing to accept deeply strange implications to get there.

With this one impulsive reach for a solution to a prickly problem, Planck had inadvertently taken a crucial step towards a profound revolution in physics. Natural phenomena like electromagnetic radiation were presumed to exist as a continuous spectrum, like the rainbow of colors extracted from sunlight by a prism. At a human scale, even at a microscopic scale directly observable by technological means, the continuity was readily apparent. Below that scale, extrapolations of the visible continuum fell apart. Now Planck's solution suggested that the radiation physicists were predicting theoretically and measuring experimentally was less like a volume knob on an old stereo, smoothly transitioning from 0 to 10 (or 11 as the case may be), and more like a channel changing dial on a television, snapping between channels with only nothingness in between.

The next shock wave came for physicists in 1905, when Albert Einstein tackled the difficulty that classical physics had with explaining the photoelectric effect. Experimentalists had observed that when metal is exposed to ultraviolet light, electrons are emitted. This is the phenomenon we rely on to make smoke detectors, digital cameras, solar panels and all manner of other technologies work. However, the wave theory of light was unable to explain some of the behaviors the experimentalists observed. Einstein used Planck's "quanta" of energy and applied it to light, proposing that it exists in discrete quantities he called photons. This seemed to contradict the work Fresnel had done to establish the wave nature of light, even while the classical wave theory from Maxwell could predict and explain so much of light's observable behaviors. Despite the paradox, Einstein's explanation produced results that matched the experimental results, the idea slowly began to take hold, and Einstein would go on to win a Nobel prize for the work in 1921.

Whether the Nobel for photons legitimized the early work of Planck and Einstein or their ideas had simply had enough time to incubate, the rest of the decade was a very busy one for quanta. In 1924 the French physicist de Broglie proposed in his doctoral thesis that the wave-particle duality was not restricted to massless photons, but that electrons, and by extension, all matter could exhibit both wave and particle behaviors. This theory suggested there was no neat solution to the questions around sub-atomic behaviors coming around the corner. In fact, the more physicists investigated the more they found confounding and mystifying behavior.

Mystifying behavior was further validated in 1925 by Werner Heisenberg, a German physicist who had been working at the University of Göttingen, where Planck had also been a professor. Summoned by Niels Bohr to Copenhagen to help work with a problem. Bohr, a giant in the field already, had created formulae that could accurately predict properties of chemicals such as the frequency of light they would emit when heated before experimental results were achieved. However, the internal logic of these predictions mystified him, and on the advice of Wolfgang Pauli, he had called on Heisenberg to help. The issue was that, in order to work, the formulae had to assume that electrons orbited their nuclei only at very precisely defined orbits and not at some arbitrary distance from the center. What's more, when excited, electrons didn't seem to ascend linearly from a lower orbit to a higher one, the way you'd expect a satellite orbiting the earth to smoothly move into larger and larger orbits by firing their jets. Instead, the electrons seemed to leap from one allowable orbit to another. The accuracy of the predictions strongly implied those assumptions were correct but Bohr had no idea why.

In June of 1925, suffering from hay fever, Heisenberg traveled to Helgoland, an island off the German coast with no trees, hence no pollen, to seek relief. There, at 3am one morning, he completed work on a hypothesis which left him shaken and disturbed.[6] Like Planck, he'd found himself desperate for ideas that would provide him with a way to make the mathematics work, and so he'd decided to taken as given that the strange leaps predicted in Bohr's equations reflected the actual situation at the subatomic scale, and had constructed tables of numbers that represented the allowed positions for each electron. In one position in the table was the electron's current orbit, in another, the orbit it was leaping to, its destination. The table or matrix mathematics was clumsy, but it worked. The initial paper written from the Helgoland insight provided not only a working model to explain Bohr's strange leaps, but also the raw material Heisenberg, Born, and Jordan Pasqual needed to create the foundational model for quantum mechanics. Bohr's orbital leaps are not arbitrary, but seem to be part of the universe's internal design.

The following year, 1926 Erwin Schrödinger, an Austrian working at the University of Zurich, delivered a generalized theory of de Broglie's wave-particle duality that also succeeded in calculating Bohr's results. Schrödinger sought to create a mathematical way to describe and predict the behavior of a particle when it acted more like a wave. The resulting equation seemed a very elegant solution, calculating the function of a wave Schrödinger calls ψ, or Psi, after the Greek letter, and it reproduces the same results as Heisenberg, without the awkward matrices. Heisenberg saw a critical flaw in the wave model for the primacy of Schrödinger's wave function in the observable behavior of waves. Traveling over a distance, a wave will arrive at a distant point a diffused version of itself, the ψ having spread out evenly over the course of its journey. An electron arriving at a location, however, arrives as a single point, a phosphor on a television screen, not a smear. Heisenberg found Schrödinger's work "repulsive," [7]and was greatly motivated to reassert his own explanation for the physical world.

One of the oddities of the matrix mathematics of Heisenberg's model was non-commutativity, in other words, the order of calculations matters. In most types of math, the terms on one side of an equation can be read right to left or left to right, it doesn't matter because "4 x 5" yields the same value as "5 x 4." However, in matrices, the order does matter. Multiplying the position of a quantum system by its velocity does not give the same answer as multiplying its velocity by its position. Because of the relationship between non-commutative pairs like position and velocity or momentum, an uncertainty exists between them, and Heisenberg realized that knowing the value of one made the value of the other attribute impossible. In 1927, he published his work in a paper called *On the Perceptual Content of Quantum Theoretical Kinematics and Mechanics,* in which he made an assertion regarding the uncertainty of the value of one of a non-commutative pair when one measures the other attribute of the pair.

Heisenberg's uncertainty principle provided a theoretical basis that proved the wave-like behavior of particles put limits on how precisely we could measure their properties. In fact, the more precisely we measured one attribute of certain related pairs, like position and momentum, or time and energy, the more uncertain the other attribute became. That is, if you tried to fix a particle's position in space, its momentum, or the product of the particle's velocity multiplied by its mass, became less and less certain. In this way the uncertainty principle integrated Schrödinger's wave function as a behavior exhibited within Heisenberg's framework of quantum mechanics. Unmeasured, an electron or a photon's attributes were undetermined, a probabilistic wave, Schrödinger's ψ. Once observed, the wave function collapsed and the particle was a particle again. Heisenberg's uncertainty principle described a dead end for the path of Rationalist scientific progress in the sense that it proved that it was impossible to precisely define all the characteristics of a given subatomic particle, and therefore to fully understand the larger system of which it was a part.

Heisenberg, by formulating probability distributions that described a particle, implied that a particle had no fixed attributes but rather were determined probabilistically when measured. Worse, Schrödinger's equation implied non-local influences on those probabilistic outcomes, meaning a quantum particle's attributes might be influenced by some other particle at an arbitrary distance. This new physics was extinguishing the Rationalist mission to trace the phenomena of the physical world and everything in it back to its root causes, its Godhead, presumably a set of equations from which one could derive all the contents of the universe and all of the interactions and ultimate fates. Humans are fond of the assumption that there is an objective reality that may be extremely difficult to accurately discern but ultimately does have a solution, an ultimate answer, a rational reason for everything that has happened and will happen. As quantum physics continued to take shape it seemed to pose a threat to the entire notion of objective truth. There did not seem to be any majestic, cosmic watch works underneath the surface of it all.

These disturbing implications became the backdrop for a major physics conference held in 1927, in Brussels. Named for the Belgian industrialist and philanthropist who provided support for the event, the Solvay conference had begun in 1911 and in 1927 was being held for the fifth time. Given the recent results from de Broglie, Heisenberg, Schrödinger and others, the implications of the emerging theory for the behavior of quanta, or quantum mechanical physics, was the central topic of debate.

A consensus was forming among a majority of the Solvay participants, who were aligning with the formal convergence of the recent work in quantum mechanics lead by Niels Bohr, Heisenberg, Max Born and others. Einstein, despite his role in discovering that light also existed as discrete quanta of energy, was convinced that quantum mechanics was incomplete or incorrect in some way. In a letter to Max Born the previous year he'd expressed his concerns with the disturbingly non-deterministic universe suggested by emerging physics, writing, "the theory says a lot, but does not really bring us any closer to the secret of the 'old one'. I, at any rate, am convinced that *He* is not playing at dice."[8] Einstein's concern connected back to the Rationalist dream of a universe that would make sense to *someone,* even if it remained beyond the abilities of any individual to understand. With perfect knowledge, that of a hypothetical supreme being, for example, every aspect of the universe should follow from some other aspect, an unending chain of cause-and-effect certainty. What quantum mechanics suggested was that no such perfect knowledge was possible, that events in the universe were, in the end, determined randomly and impossible to predict.

The participants at Solvay, seen in *Figure 3.4*, split into two factions. A group of the quantum skeptics, led by Einstein, faced off with Bohr and the others who were stitching together the work of the previous 20 years into a coherent formalism.

Figure 3.4: The participants of the Solvay Conference, 1927.

This set of axioms included a belief that the wave-particle duality applied to quantum particles such as electrons and photons, which could behave as waves in certain contexts. While behaving as a wave, a particle can be in superposition of states, a concept that defies human intuition and most analogies used to describe it. If a particle has an attribute that has two possible states, let's call them "up" and "down," while in superposition its state will be some probabilistic combination of both states. It is not accurate to say it is both up and down simultaneously, nor that it is some deterministic mix, for example 30% up and 70% down. In fact the probabilistic parts, the "amounts" of up and down, are unknowable. Once the particle is measured, its wave function will "collapse," resolving into a single state, up or down, no longer in superposition.

As Heisenberg had previously discovered, certain pairs of attributes of quantum particles such as position and momentum can't both be known at the same time, and the more accurately you measure one, the less precisely you know the other. Finally, Bohr's contribution to all this quantum weirdness was the concept of complementarity, which held that certain phenomena such as the wave-particle duality, are determined by the experimental setup. Put in terms of

Young's experiment, if light is shone through a single slit the photons will behave as particles, and if it is shone through two slits, they behave as waves. If one attempts to measure which slit each individual electron goes through, the wave behavior will disappear and photons will behave as particles again. In terms of our explanation of superposition above, when in superposition the photons have a probability density to pass through one slit or another, and when measured their wave function collapses and they behave like particles again.

After Solvay, the two camps continued to develop their respective positions, with the quantum mechanics theory coalescing at Bohr's institute at the University of Copenhagen, and coined the "Copenhagen interpretation" by Heisenberg in 1935. Einstein for his part published work[9] with Boris Podolsky and Nathan Rosen which cast doubts upon quantum mechanics' completeness. The implications of the Copenhagen interpretation included the basis of the thought experiment that came to be known as the Einstein-Podolsky-Rosen, or EPR paradox. The EPR paradox posited a pair of particles prepared in such a way that, if one were to measure an attribute such as momentum of one particle, the momentum of the other would also be known instantaneously, even if the particles were separated by an arbitrary distance. This, the EPR paradox contended, would violate the theory of relativity, which held that nothing could travel faster than the speed of light. Since the measurement of one entangled particle would determine the collapsed state of the other particle regardless of the distance separating them, the implication was that the information was conveyed from one to the other faster than light itself could travel. In another letter to Max Born, Einstein laid out his objections to quantum mechanics, writing, "...physics should represent a reality in time and space, free from spooky action at a distance."[10] The crux of the problem for Einstein was "non-locality," or the action of one particle on another particle at any distance from the first. The paper posited the existence of "hidden variables" not known to quantum theory that would provide a rational resolution to the paradox.

Einstein wasn't the only pioneer of quantum theory who felt it must be incomplete. Erwin Schrödinger, whose wave function was so crucial to the Copenhagen interpretation, created his own thought experiment, also in 1935, in an attempt to prove the absurdities that quantum mechanics implied were possible. Writing in response to the EPR paradox, Schrödinger's paper, "[11]The present situation in quantum mechanics," imagined a cat in a steel box along with a small amount of radioactive material and a device that detects radiation, a Geiger counter. If the Geiger counter detects a particle emitted by the radioactive material, it will trigger a mechanism that breaks open a vial of poison, and the cat will die. Since the emitting of radiation is modeled using the wave function, the Copenhagen interpretation holds that the wave will collapse when "measured." Schrödinger's cat would not "collapse" into a state of alive or dead until an observer opened the box and determined the wave function collapse.

Before the box was opened the cat would be both alive and dead in some probabilistic combination.

Various responses from Bohr and others suggested fallacies and alternative perspectives on both the EPR paradox and Schrödinger's cat, and the Copenhagen interpretation survived and even thrived. The definitive blow came decades later, when Irish physicist John Bell devised a theorem in 1964 that could distinguish experimentally between quantum mechanics and any non-quantum hidden-variable theory of the kind Einstein had hoped for. In the 1980s, Alain Aspect and his colleagues in Paris carried out the experiments, and the results were unambiguous: quantum mechanics was right, Einstein's hidden variables were not there, and "spooky action at a distance" was real. It remains one of the most dramatic confirmations in the history of science. The long, gradual erosion of the primacy of classical physics was now complete. The Copenhagen interpretation remains to this day the dominant understanding of the workings of our universe. Classicism and Rationalism had run their course and been ultimately defeated by uncertainty and probabilities.

However, it's important to note that the Copenhagen interpretation, while historically significant and still widely taught, is just one competing view of quantum mechanics, and there are others with their own supporters and detractors. The debates from Solvay were just the beginning, with the Copenhagen interpretation setting the stage for a broad set of theoretical and experimental explorations yet to come. After the watershed year for quantum physics in 1935, its esoteric pursuits had to take a back seat to more pragmatic concerns. The world was at war, again, and scientists were being called on to contribute more directly and profoundly than in any prior conflict.

References

1. Andrew Robinson, The Last Man Who Knew Everything: Thomas Young. Cambridge, UK:Open Book Publishers, 2023, `https://doi.org/10.11647/OBP.0344`
2. James Clerk Maxwell, "A Dynamical Theory of the Electromagnetic Field," Philosophical Transactions of the Royal Society of London 155 (1865): 459-512, `https://doi.org/10.1098/rstl.1865.0008`.
3. "Fresnel's Wave Theory." Field Guide to Polarization. SPIE, n.d. `https://doi.org/10.1117/3.626141.p5`.
4. Not to be confused with the album of the same name by the American band Trotsky Icepick, released in 1991.
5. Max Planck, letter to Arnold Sommerfeld, October 27, 1930, Max Planck Gesellschaft Archiv, Berlin, Germany, cited in John D. Norton, "Max Planck's Derivation of the Black-Body Radiation Formula," Archive for History of Exact Sciences 45, no. 1 (1992): 17-94.

6. Heisenberg, Werner, Der Teil und das Ganze (Munich: Piper, 1969)
7. Manjit Kumar, Quantum: Einstein, Bohr and the Great Debate about the Nature of Reality (London: Icon Books, 2008),155
8. Albert Einstein, letter to Max Born, 4 December 1926, in The Born-Einstein Letters (New York: Macmillan, 1971), 91.
9. Einstein, Albert; Podolsky, Boris; Rosen, Nathan (May 15, 1935). "Can Quantum-Mechanical Description of Physical Reality Be Considered Complete?". Physical Review. Princeton, New Jersey: Institute for Advanced Study. 47 (10): 777–780. Bibcode: 1935PhRv...47..777E. doi:10.1103/PhysRev.47.777
10. Albert Einstein, letter to Max Born, 4 December 1926, in The Born-Einstein Letters (New York: Macmillan, 1971), 158
11. Schrödinger, Erwin (November 1935). "Die gegenwärtige Situation in der Quantenmechanik (The present situation in quantum mechanics)". Naturwissenschaften. 23 (48): 807–812. Bibcode:1935NW.....23..807S. doi:10.1007/BF01491891. S2CID 206795705.

4

Mid-Century Classical

In natural science, Nature has given us a world and we're just to discover its laws. In computers, we can stuff laws into it and create a world.

— Alan Kay

Information gets theoretical and applied

While the debate over the completeness and interpretation of quantum mechanics were captivating the attention of the physics community, the rest of the world was preoccupied with a very different topic in 1935. The previous year had seen Hitler declare himself both Chancellor and President of Germany, a self-styled "Führer," and in 1935 declared the country would begin re-arming itself, violating the Treaty of Versailles. Mussolini invaded Ethiopia in defiance of the League of Nations, revealing its powerlessness to stop aggressions, and the world braced itself for war. In the same year, Robert Watson-Watt, a Scottish physicist, demonstrated a method that used radio waves to detect a British bomber at a significant distance, proving the feasibility of a system called "radio detection and ranging," or Radar. Daydreaming about the behavior of subatomic particles might be fascinating, but it did nothing for national security. Physicists turned from the theoretical to the practical and began to contribute to their nations' readiness for war.

As the world descended into conflict, scientists continued to work on practical applications of the theoretical advances of the prior decades. Radar was improved through the collective work of allied scientists and engineers delivering improvements like the cavity magnetron, which used higher frequency microwaves to detect aircraft, greatly extending range over radio waves. Wireless communication brought greater agility to the battlefield, but it also exposed a nation's plans and strategies to a new vulnerability. Radio transmissions were routinely intercepted, so the warring nations began to increase the sophistication of their encryption methods through the use of complex mechanical devices like the German Enigma machine.

Even capturing an Enigma device, as the British Royal Navy did in 1941, was not enough to break the German code, as the elaborate scrambling scheme had so many variations based on the seed chosen before encryption that the best mathematicians and puzzle solvers in the UK could not defeat it in a timely fashion. The UK launched two separate secret projects to overcome the German code, both at Bletchley Park. One effort sought to build machines using mechanical relays, called Bombes, that would use the captured knowledge of the Enigma machine to accelerate the breaking of an encrypted message. Critical to the development of the Bombes was a young mathematician named Alan Turing. While the Bombe was not a computer *per se,* since it was only designed to carry out a narrow set of computations and was not programmable, the other project at Bletchley Park was. Known as "Colossus" and developed by engineer Tommy Flowers, it was the first electronic, programmable, digital computing device, and it was used to crack an even more complex German code, the Lorenz cipher.

Flowers, son of a bricklayer, took evening classes at University of London to earn a degree in electrical engineering, then joined the telecommunications branch of the Royal Post Office. When Turing asked his director in 1941 for help with the Bombe, he was introduced to Flowers. The Bombe used mechanical relays, and Turing was exploring how vacuum tubes, "valves" as they were known in the UK, might improve his machine. Mechanical relays are electromagnetic switches, where an electrical current is used to physically open and close a circuit — a light switch with a magnet taking the place of a human finger to flick it up and down. Vacuum tubes had no moving parts, instead using current in one lead in the tube to induce current to flow between the other two leads. In effect, tubes moved the wall switch's function *inside* the light bulb. Flowers had designed and built telephone networks entirely using electronic valves instead of electro-mechanical relays, and provided some advice. Turing was impressed by Flowers and encouraged him to develop his ideas further. Using his own money to develop the prototype, he eventually convinced the government his system could work, and by December 1943, the Colossus Mark I with its 1800 vacuum tubes was operational at Bletchley Park. A restored Colossus Mark I is seen in *Figure 4.1*, below.

Figure 4.1: A rebuilt Colossus breaks a cipher at Bletchley Park in 2006. By MaltaGC, CC BY-SA 3.0, https://commons.wikimedia.org/w/index.php?curid=2657152

The Colossus system was programmable using a variety of switches, patch cables, with data entered via a paper tape input, all of which in combination could provide the inputs the machine needed to decrypt the Lorenz cipher message. Most notably, the input on the tape consisted of a series of holes that represented the encrypted message in binary form. While it was not a general-purpose computer, the Colossus could be said to be the realization of Charles Babbage's Analytical Engine. Conceived of in the 1830s, the Engine, as seen in *Figure 4.2*, was a mechanical computer designed to take punched cards as input for computations performed by a central "mill" with results delivered into a memory store. The Analytical Engine was never built, but in the sense that it was designed as a general-purpose computer, it could be said to have been more sophisticated than the Colossus. Babbage had initially designed and prototyped a simpler machine, the difference engine, which was more calculator than computer, and not general purpose.

Figure 4.2: Trial model of Babbage's Analytical Engine, on display at the Science Museum in London, UK

His collaborator from the early 1840s onward, Ada Lovelace, was the first to recognize the potential of a general-purpose computer for more than just math, writing it

> *...might act upon other things besides number, were objects found whose mutual fundamental relations could be expressed by those of the abstract science of operations, and which should be also susceptible of adaptations to the action of the operating notation and mechanism of the engine...Supposing, for instance, that the fundamental relations of pitched sounds in the science of harmony and of musical composition were susceptible of such expression and adaptations, the engine might compose elaborate and scientific pieces of music of any degree of complexity or extent.*[1]

Babbage and Lovelace form the canonical example of the inseparable dependence of computer engineers on computer scientists, and vice versa, mirroring the same relationship dynamic between experimental and theoretical physicists. The collaboration between experimentalists and theorists is sometimes called "co-design," a term that attempts to capture the symbiotic and interdependent nature of the work done by both camps. In the middle of the 20th century, co-design was a vital ingredient in the astonishing pace of progress in computation. Before Alan Turing was helping design the Bombes at Bletchley Park he'd been a gifted mathematician, who in 1937 had published what has been called "easily the most influential math paper in history,"[2] where he laid out the foundation for the entire field of computation.

Titled *On Computable Numbers, with an Application to the Entscheidungsproblem*[3], the paper was Turing's attempt to address a problem posed by the mathematician David Hilbert, the "Entscheidungsproblem," or "decision problem." Hilbert had asked in 1928 if there were a standard method or process which can be applied to a mathematical proposition which gives a decision as to the truth or falsehood of the proposition?[4] What Hilbert was getting at was, in turn, a sharpening of a question posed by the staggering genius of Gottfried Leibniz, who, in the 17th century, built a calculating machine that predated Babbage by two hundred years, and wondered if such a machine could be built to not only solve all math problems, but also to identify those that had no solution.

Turing approached the problem through the theoretical work of another mathematician and logician, Kurt Gödel, whose incompleteness theory in 1930 laid out a proof that any set of logical axioms will have limits to what it can prove. Extending that line of thinking to Hilbert's question, Turing's 1937 paper described a hypothetical machine which read instructions and data from a tape and carried out operations based on the input. Having described the idea of a digital computer as a culmination of a centuries-long conversation among theorists, Turing found himself just a few years later involved in the efforts to practically realize the dream of a universal computing machine.

Before the war efforts, Turing's studies brought him to Princeton university in the U.S., where in 1936 he began his PhD with Alonzo Church as his advisor. Church had also been working on the problem of computability, and in fact had published a paper very much along the lines of Turing's work. Their parallel efforts are recognized today as the Church-Turing theorem.

Flowers and Turing were not the only ones working on programmable computers. In fact, under the Nazis, the Germans had not only developed the Enigma and the Lorenz code machines, limited purpose computers in their own right, but had also built, in 1941, the world's first automatic, digital, programmable computer. Built by Konrad Zuse, who, like Flowers, was an engineer, the Z3 system used a punched tape to load instructions and was able to perform floating point arithmetic. Meant to speed the calculations required for aircraft design, the system failed to capture the imagination of the Nazis, and when it was destroyed in allied bombing raids in 1943, thankfully, there was no effort to replace it.

Meanwhile, in the US, similar scientific and engineering efforts were afoot. At MIT, Norbert Wiener, an American-born child of Jewish immigrants, a prodigy, a philosopher and mathematician who had studied with David Hilbert, worked on systems to control anti-aircraft guns automatically, using analog electronic means to track and predict the path of enemy planes. The U.S. army set up the Signal Intelligence Service as a sort of counterpart to Bletchley Park, with a focus on breaking the codes used by the Japanese in the pacific theater. And at Bell Labs, a researcher by the name of Claude Shannon was working on theoretical understanding of information.

Working on analog computing devices during his graduate studies in electrical engineering at MIT in the 30s, Shannon had drawn on the work of George Boole to create switching circuits that allowed the simplification of the electromechanical relays on the machine. This, like the experience of Tommy Flowers, had applications in the design and operation of telephone networks. In fact, Shannon had, as a child, used a barbed wire fence as a telegraph wire to communicate with a friend. Shannon's work after earning his PhD at MIT was more theoretical in nature. He joined Bell Labs in 1941 to work on fire control systems along the lines of Wiener's cybernetics work, and on cryptographic systems.

His involvement in code breaking brought him in contact with Alan Turing, who in 1943 had been sent to the U.S. to share code breaking techniques from Bletchley Park with the Navy. During a visit to Bell Labs, Turing and Shannon met in the cafeteria, and Turing walked Shannon through his paper *On Computable Numbers,* which had a huge impact on Shannon. Much of his work during the war at Bell Labs culminated in his 1948 paper A Mathematical Theory of Communication, where among his many contributions he coined the word "bit" to mean a binary digit with a value of 0 or 1. His theoretical work was to lay the foundation for information theory and much of digital communications networks for decades to come.

The work of Tommy Flowers to build Colossus had its doppelgänger in the U.S. as well. The U.S. Army, wanting a faster way to calculate the large number of differential equations required for artillery firing tables, used to aim artillery barrages, funded an effort at the University of Pennsylvania to build a computer. John Mauchly and J. Prosper Eckert began building ENIAC, or Electronic Numerical Integrator and Computer, starting in 1943. Inspired by the design of Vannevar Bush's Differential Analyzer, for which Shannon had created his logical circuits, the ENIAC was, like Colossus, a programmable, general-purpose computer that used vacuum tubes for gates.

With the problems of breaking enemy codes largely solved, the champions of computing began to look to other challenges which their increasingly capable machines could tackle. Even before ENIAC was fully operational, the idea of using it for the Manhattan project was raised. The project to build the atomic bomb was enormously complex, requiring the work of a small army of human "computers," mostly women, to carry out the mathematics. Frustrated by the slow pace of progress, Richard Feynman, then managing a group of computers at Los Alamos for the Manhattan project, came up with an innovative approach to breaking down a large problem into smaller ones that could be solved in parallel, a technique that closely resembles modern approaches to parallel digital computing.

Another key figure at Los Alamos, John Von Neumann, was well aware of both the ENIAC effort and Feynman's and other's efforts to make human computers more efficient. Von Neumann had been at Princeton's Institute of Advanced Studies (IAS) since shortly after its creation in the early 30s. Von Neumann began lining up support from Princeton and secured funding from the Office of Naval Research and the Atomic Energy Commission before starting to build his machine in 1946.

With the war over, von Neumann's efforts were helped by the availability of surplus components such as vacuum tubes, which, like ENIAC or Colossus, the IAS machine used for logical gates. Memory was implemented using a novel approach where the cathode ray tubes were scanned continuously and what eventually would be called a pixel's state was either read out or changed. The IAS team took advantage of the fact that the glow of an activated spot on the tube would persist between scans of the electron gun and that the gun itself could 'sense' when a spot was illuminated or not. Forty tubes were used to store an array of 1024 "words," each 40 bits long, for a total 40,960 bits of memory. The IAS machine was also innovative in the way it loaded its program from punched tape into the same memory space as the data, creating an instance of Turing's tape abstraction in those tubes.

The team was challenged by the quality of the surplus components, especially the 6J6 vacuum tubes, which they ended up using about 2000 of in the machine. Faulty tubes were easy to detect with startup diagnostics, and could be replaced. The real issue was the tubes operating out-of-spec, introducing subtle errors over time to the machine's operations. Ingenuity and engineered redundancy lead them to solutions to their problems, all of which have left a mark present on contemporary computers.

Figure 4.3: John von Neumann with the system built at the Princeton Institute for Advanced Studies. Shelby White and Leon Levy Archives Center.

What was potentially more challenging for von Neumann was to maintain support from the sources of funding in the new post-war context. His involvement in the Manhattan project allowed him to create solid arguments for the use of his computer to continue research into nuclear weapons, a topic that seemed germane given the newly simmering rivalry with Soviet Russia. For those who wanted a post-war focus on peace and prosperity, von Neumann could emphasize the role of computation in the development of nuclear reactors for energy generation, beckoning to the promise of the "atomic age." More broadly, von Neumann seemed to have an instinct for the ability of digital computers to aide in scientific inquiry, especially through the use of simulation or modeling.

In an effort to demonstrate the potential for computers to simulate natural processes other than the compression and blast of a nuclear weapon, von Neumann brought a leading meteorologist, Jule Charney, to IAS, where he built the first computer models to simulate weather patterns. Another researcher invited to use the IAS machine was Nils Barracelli, an Italian-Norwegian scientist who pioneered the field of artificial life, simulating mutations and evolutionary pressure among life forms simulated in the vacuum and cathode ray tubes of the IAS machine.

The researchers and engineers working around the IAS machine created the Monte Carlo method, an immensely powerful statistical approach that creates an approximate model of a complex system using random sampling. Invented by Stan Ulam, the Monte Carlo method remains tremendously useful to this day in physics, engineering, finance, and artificial intelligence.

Though the IAS machine was decommissioned in 1958, just 13 years after the beginning of the project, its impact on the development of classical computing was profound. Von Neumann seemed to have a keen sense for the most powerful theoretical and experimental ideas from the scientists, mathematicians and engineers contributing to the war effort, which he drew on to powerful results. Most importantly he demonstrated a very open attitude towards the creations of the team at the IAS, happily running off copies of the machine's specification to anyone who asked. As a result of being shared freely in this manner, there were fifteen documented "clones" of the IAS machine built at other institutions around the world.

In contrast, Eckert and Mauchly, who led the ENIAC and UNIVAC projects at the University of Pennsylvania, were quite territorial about their work. They complained that Von Neumann and others were "snooping around" their work, and in 1946 formed the Eckert-Mauchly Computer Corporation and began filing patents to protect what they viewed as their intellectual property. The legal fight over these foundational patents took decades to play out, first being granted to Eckert and Mauchly in 1964 and finally being overturned in the early seventies. In the meantime, the IAS clone devices, with derivative and humorous names like "MANIAC" at Los Alamos, or "Johnniac" at the RAND Corporation, had a profound and lasting influence on information technology. In what might be the first example of the power of an "open source" strategy in computing, Von Neumann's influence is apparent to this day in classical computing architectures in everything from a web server to the smart phone in your pocket, while the ENIAC architecture is lost to the ages.

However, if the computational demands that provided the motivation and funding for the Bletchley Park Bombes, Colossus, ENIAC and the IAS system had remained their sole motivation, the momentum gained in the 1940s may have dissipated in the post-war era. Instead it seems that every researcher who got their hands on the machines or even read about the work that was being carried out on them was able to imagine more uses for the technology, and dreams of a digital future continued to gain steam. In fact, immediately after WWII came one of the critical defining moments in the history of classical computing: the invention of the transistor.

The devices built during the war had transitioned from mechanical relays to vacuum tubes, which had brought huge gains in speed and scale. However, like a light bulb, tubes were prone to physical failure as their components heated and cooled repeatedly through their duty cycle. Thankfully, as interest in machines that used dynamic circuits to solve Boolean logical problems surged after the war, physicists were hard at work building out our working knowledge of the universe in ways that could improve upon vacuum tubes. In particular, John Bardeen and Walter Brattain, two researchers at Bell Labs in New Jersey, were exploring the characteristics of materials known as "semiconductors." The term was invented in the WWII era to describe a category of materials first identified at the end of the 19th century which had a mix of characteristics of conductors like copper and insulators like glass. Of particular interest to researchers like Bardeen and Brattain was the ways in which semiconductors' behavior could be manipulated and controlled. Their experiments resulted in a demonstration of a Germanium strip with two electrodes which, when a small current was applied to one end, would produce a much-amplified current from the other end.

Others at Bell Labs such as William Shockley built upon this experiment and, in 1947, created what they called the junction transistor. Consisting of three layers of semiconducting material sandwiched together, a junction transistor was capable of powerfully amplifying electrical signals, which would lend its name to the new, cheap and compact "transistor radio." More importantly, a junction transistor could act as a digital switch — when current was applied to the middle section of the junction, current could flow from one end to the other. Turn off the current to the middle, and the current could not flow.

The typical vacuum tube used in computing devices could last a few thousand hours in operation before its filaments burnt, its vacuum failed, or some other mechanical flaw ended its useful existence. With no delicate components, the "solid state" transistor far exceeded the lifespan of tubes, with some transistors from the 50s and 60s in operation to this day. A huge barrier to building complex, useful computational devices that didn't require the continuous application of specialized skills to maintain had been removed.

Bell Labs' breakthrough ushered in the electronic age due to the simplicity, low cost, and durability of the transistor. The earliest pioneers of the vacuum tube era of computing such as Remington-Rand, who had bought the UNIVAC from Mauchly and Eckert, IBM, with its model 701 "Defense Calculator," or Ferranti and English Electric in the UK, all began transitioning to solid state machines. However, there were still challenges to building robust universal computers capable of productive work at scale. Transistor-based machines had to be assembled by hand, with workers soldering each tiny component to circuit boards. Solder joints are fallible and a single cracked joint could cause the machine to immediately stop. Even the most skilled workers with the best materials and the most controlled environment have some probability to create a faulty joint. Multiply that probability by a sufficiently large number of components soldered to a board, and the overall chance of failure will rise. The largest transistor-based mainframes by the late 50s, such as the IBM 7090, had 50,000 components.

This challenge was the dominant theme of the 1950s, with huge amounts of energy and R&D budget focused on potential solutions, none of which made much of a dent. Approaches like MERA, or Modular Electronics for Radar Applications, funded by the U.S. Air Force, which tried to establish a Lego-brick like system of electronic components that snapped together typified the attempts to solve the problem.

Then, at the very end of the decade, researchers at Texas Instruments and Fairchild Electronics realized at almost the exact same time that the solid state of transistors meant they could be "printed" on a monolithic sheet of semiconducting material. By 1962 the defense industry was using integrated circuits for the guidance system of the Minuteman missile, and in 1964 IBM launched the System/360, with a combination of printed resistors and soldered transistors. The electronic age began to give way to computers, and integrated circuits paved the way to the microprocessor.

In the early, heady days of the computer age, it seemed as though these machines were capable of anything, with reach generation yielding ever more powerful devices. Gordon Moore's observation that the number of components on a chip had been doubling roughly every year was originally made in 1965, looking back over the previous several years of progress, and though the frequency had stretched to every 2 or so years by 2020, it's been an astonishingly constant rate of innovation.

These miracles of science and engineering, however, were imagined, designed, and built in the context of Shannon and Turing's classical information theory. While extremely powerful, the reliance on Boolean logic creates a weakness, with certain types of problems demanding exponentially more computing power to solve than even our most powerful machines can deliver. This shortcoming may be solved by the introduction of quantum mechanical concepts into information theory, if we can invent the machines to do it.

References

1. Lovelace, Ada; Menabrea, Luigi (1842). "Sketch of the Analytical Engine invented by Charles Babbage Esq". Scientific Memoirs. Richard Taylor: 694.
2. Avi Wigderson (2019). Mathematics and Computation. Princeton University Press. p. 15. ISBN 978-0-691-18913-0.
3. Turing, A.M. (1937). "On Computable Numbers, with an Application to the Entscheidungsproblem". Proceedings of the London Mathematical Society. 2. 42 (1): 230–65. doi: HYPERLINK "`https://doi.org/10.1112/plms/s2-42.1.230`"`10.1112/plms/s2-42.1.230`. S2CID HYPERLINK "`https://api.semanticscholar.org/CorpusID:73712`"73712.
4. David Hilbert and Wilhelm Ackermann. Grundzüge der Theoretischen Logik. Springer, Berlin, Germany, 1928. English translation: David Hilbert and Wilhelm Ackermann. Principles of Mathematical Logic. AMS Chelsea Publishing, Providence, Rhode Island, USA, 1950

Get this book's PDF version and more

5

Gedankenexperiment

Ordinary classical information, such as one finds in a book, can be copied at will and is not disturbed by reading it. Quantum Information is more like the information in a dream. Recalling a dream and describing it to someone else inevitably changes your memory of it. But unlike dreams, quantum information obeys well-known laws.

— *Charles H. Bennet*[1]

Quantum brings its weirdness to information technology

Turing's and Shannon's information theory had been turned into a practical technology by teams working at Princeton and the University of Pennsylvania in order to aid in the calculations needed to make quantum mechanics into a terrible weapon. After the war, the focus for those building computers was for applications of a less destructive nature, and so fields of information science and physics headed off, for the most part, in different directions. Quantum physics was there to help create the semiconductor-based integrated circuits and microprocessors, and the create the lasers in CD-ROMs and fiber optics, but it had no role in algorithms and applications. That separation, it turned out, would not last long.

It is not usually possible when tracing the development of science or technology to find a tidy origin story for a breakthrough. However, this is not the case with quantum computing. The work of Rolf Landauer, a physicist who had joined IBM's research division in 1952, can be seen as a clear point of origin for the field. Landauer[2], a Harvard-trained physicist whose family had left Germany to escape Nazism, was interested in the physics of computation, especially the thermodynamic limits of computational operations. This was motivated at first for practical purposes, to understand how to better manage waste heat in computer systems. His interests would come to influence others in the field, inspiring new work or cross-pollinating with other scientists' ideas until culminating in an integration of quantum mechanics into information theory to create a new "quantum information science," or QIS.

Landauer's own inspiration arose in part from another physicist, one we encountered in chapter three, James Clerk Maxwell, the creator of classical wave theory of electromagnetism. Asked for feedback on a manuscript in 1867, Maxwell created an imaginary scenario[3] in which two chambers are filled with gas and a passage connects one to the other. If the gas in one chamber were hotter than the other, the second law of thermodynamics dictates that the hotter atoms would tend to migrate towards the cooler chamber and vice versa, until the two chambers had equalized. However, Maxwell imagines a "finite being," seen in *Figure 5.1,* who can close the passage between the two chambers at will, and who also possesses perfect knowledge of the energy, position and momentum of each molecule of gas in each chamber. When a molecule approaches the passage, if its energy is higher than the average of all the molecules, the agent allows the molecule to pass into the other chamber. If its energy is lower, the passage is blocked and the molecule stays in the chamber. Over time, one chamber will become hotter and the other cooler, decreasing entropy and violating the second law of thermodynamics.

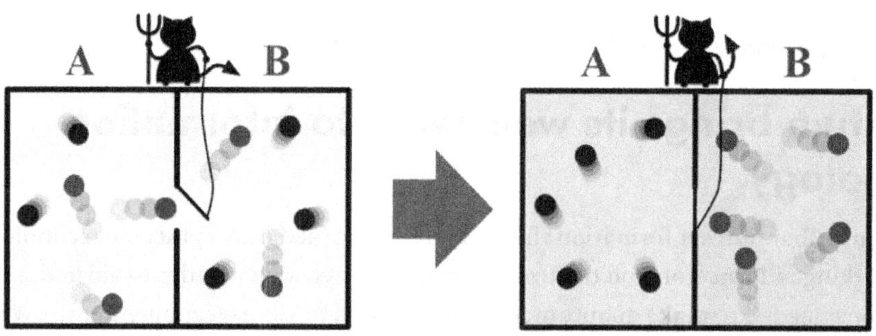

Figure 5.1: Maxwell's demon by User:Htkym - Own work, CC BY 2.5, https://commons.wikimedia.org/w/index.php?curid=1625737

The being was later referred to as "Maxwell's demon" by Lord Kelvin, a name that stuck despite the fact that the very religious Maxwell would never think of including a demon in his thought experiment. These "what ifs" are a powerful tool in physics and are often referred to by the German word *gedankenexperiment,* at least since its use by Hans Christian Ørsted in his 1811 work "First Introduction to General Physics."[4]

Maxwell's demon seems to be the type of idea that provokes irritation and a desire to explain why it is wrong in nearly everyone who encountered it. To many it seemed self-evident that the demon's closing and opening of the connection between the two chambers must require work, as would acquiring information about the molecules to make the decision about that connection. However, none of the irritated responses generated solid theoretical rationales, until Landauer's paper *Irreversibility and Heat Generation in the Computing Process*[5], published in 1961. His argument was that some of the logical operations carried out by computing machines were irreversible — once executed there was no way to reverse them. This logical irreversibility, Landauer reasoned, was tied to a physical irreversibility, an increase in entropy of the system measurable by an increase in heat. Erasing a bit, the smallest unit of information, would incur a minimum energy expenditure, even in a perfectly efficient computing device. Landauer wrote that "computer designers and even theorists of computing ought never to lose sight of the fact that every bit must be embodied and every logic operation accomplished by some real, physical apparatus."[6] Information is intrinsically physical, as much as part of nature as mass or velocity.

Before Landauer, the physics of information was not rigorously studied, and the field operated with a rough approximation of heat generated by computation tasks. His work proved stimulating to the scientific community by widening the scope of physical science to include information, but also through his stipulation that only an irreversible change of information consumed energy. This implied that there may be a way to compute without consuming any energy at all, if one's operations were all reversible.

Aside from the practical implications of better managing energy consumption and heat output, reversibility hinted at deeper implications for computing. When we process information, whether in our heads on in classical computers, we may assume that the operations are irreversible. Logical operations in a computer are called "gates" because conceptually they can be thought of as something the information, represented in classical computers as electrical current can be controlled by, in much the way Maxwell's demon open and closed the opening between the two chambers, controlling ingress and egress. The computer's gate operations evaluate the information and, depending on its value and the nature of logical operation, allow it to pass through or does not. Some of these gate operations include the erasure of information and are irreversible. The "AND" gate performs a Boolean logical operation which takes two inputs and, if both inputs have the logical value 1, will output 1. If either input has the value 0, the output will be 0. Evaluating the output of an AND gate which produces 0, one has no way to know which of the inputs was 1 and which was 0, and therefore no way to reverse the operation.

However, if all the operations performed to arrive at a result were reversible, the computation could be run backwards to its original inputs. Some logical gates were reversible, such as the NOT gate which simply inverts the value of a bit. If the output of a NOT gate is 0, it is certain that the input was 1, and so the operation can be reversed. In the early sixties, however, a set of reversible gates which were universal, capable of carrying out any possible computation, did not exist.

The idea of reversibility proved to be irresistible to Charlie Bennett. Bennett was a PhD student at Harvard who happened to attend a talk by Landauer. Bennett had been studying chemistry but the Landauer's concept of reversible computation had a profound impact on him, he switched to studying physics, and ended up at IBM Research with Landauer.

A few others were also noodling on this question. Konrad Zuse, who had built one of the earliest programmable computers in 1938, the Z1, wrote briefly in his 1969 book *Calculating Space*[7] on the significance of reversibility but didn't take his thinking much further. Another was the largely self-taught polymath Ed Fredkin who at MIT was playing with the idea of a computer made of a frictionless billiards table. If the information could be encoded in the position of the balls, with logical operations on their information formalized as a series of collisions and bounces and trajectories, Friedkin believed a reversible computer could be created. However, "frictionless" is the kind of assumption that tends to limit thought experiments to imagination alone — there certainly isn't any easy way to eliminate friction on a billiards table! Of course, there is a scale at which something very like frictionless operations do exist — the subatomic world of quantum mechanics, where Fredkin's thought experiment would eventually find its natural home.

Fredkin's thinking did result in creating the "controlled swap" logical operation, also known as a Fredkin gate, in 1972. Interestingly, sources for the original date of the billiard ball computer and the Fredkin, or controlled swap (CSWAP) gate, are hard to come by, possibly because Fredkin didn't follow a standard academic path in his career and had less pressure to document his scientific work. In any event, Fredkin did publish a paper in 1981 with Toffoli called Conservative Logic in which they laid out both the billiard ball computer and CSWAP gates.[8]. The CSWAP is a logical operation involving three bits that is both universal and preserved the number of 1's and 0's from the input through to the output. Preserving the balance of the input values in the output allowed for reversing the operation, and its universality meant that it could be used for any conceivable logical operation or computation. Fredkin gates were the first theoretical proof that reversible logical computation was possible.

The very next year, Charlie Bennett published his paper Logical Reversibility of Computation[9], which made the case that a universal computer could not only be made reversible, it could also be made "thermodynamically reversible," meaning that, since no information was destroyed via irreversible operation, no energy would be lost. In the decade since Landauer's principle was published, Zuse, Fredkin, Bennett and others had completed the theoretical foundation for logical and physical reversible computation. All that was lacking was any idea of how one could build a device that could realize the theory. The means of computation known in the early 1970s seemed very far from the kind of efficiency that reversibility would require.

However, another thought experiment would soon serve to motivate greater interest in the subject. In the years before Bennett published his reversibility paper, he had been talking with a friend of his, Stephen Wiesner. The son of Jerome Wiesner, science advisor to the Kennedy and Johnson administrations and president of MIT, Stephen was a PhD student at Columbia during the tumultuous convergence of the civil rights movement, the rise of feminism, the war in Vietnam, and the rest of the emerging baby boomer cultural wave. During student strikes in 1968 Wiesner had hit upon an odd idea. Using what he had learned about quantum mechanical principles, he dreamt up "quantum money" that could not be counterfeited. Wiesner's quantum money encoded the serial number of a bank note in the polarity of a stream of photons, which, being quantum in nature, could exist in both wave and particle form. In transit they behaved as waves and carried the information, but if someone tried to read the serial number by measuring their polarity that coherent wave state would "collapse," destroying the information and making it impossible to create a copy of the bill.

Wiesner also came up with an idea he called conjugate coding, which took advantage of some other aspects of weird quantum behaviors to convey information, and can make a credible claim to being the first proposal[10] that quantum information's idiosyncrasies could actually be useful. In addition to leveraging the impossibility of measuring quantum information without changing it in an obvious way, Wiesner also pointed out that two quantum bits could be used to carry three bits of classical information. These two insights would, in the coming years and decades, come to be known as the "no-cloning theorem" and "hyper dense encoding," but in 1970 they were so far ahead of their time as to sound unhinged. Wiesner wrote up his ideas in a paper that, according to colleagues, was submitted to the IEEE Transactions on Information Theory, and was rejected.

When Wiesner spent time with Bennett in 1970, he told him about his ideas, with Bennett feverishly taking notes, following along with Wiesner's logic. At the top of the page of notes pictured in *Figure 5.2*, Bennet wrote "quantum information theory," what may be the first known use of the phrase.

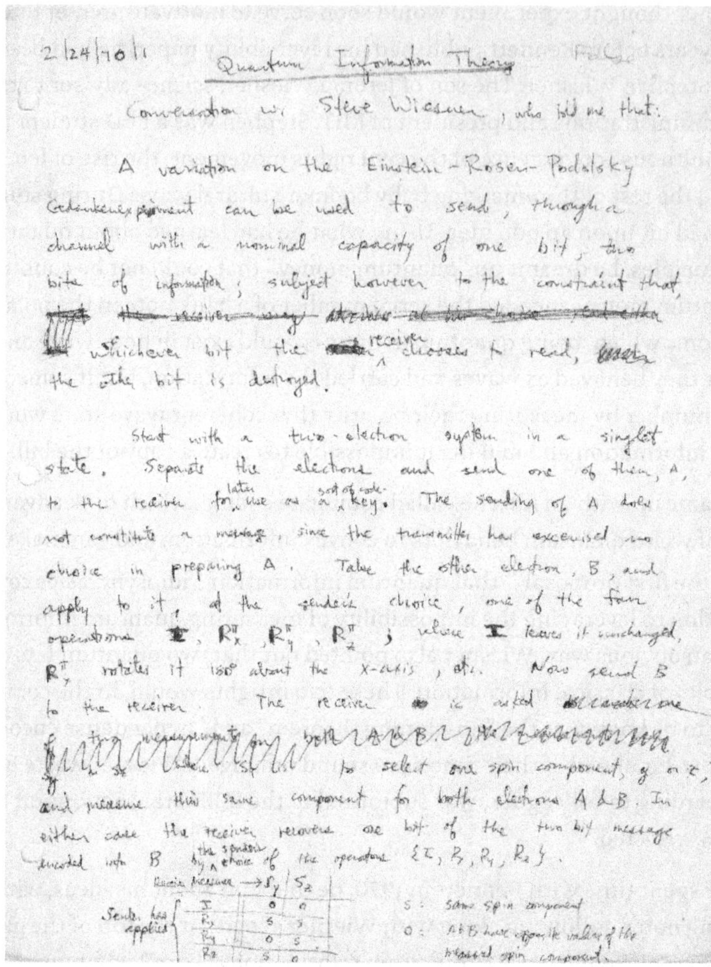

2/24/70

Quantum Information Theory

Conversation w. Steve Wiesner, who told me that:

A variation on the Einstein-Rosen-Podolsky Gedankenexperiment can be used to send, through a channel with a nominal capacity of one bit, two bits of information; subject however to the constraint that whichever bit the receiver chooses to read, the other bit is destroyed.

Start with a two-electron system in a singlet state. Separate the electrons and send one of them, A, to the receiver for later use as a sort of code key. The sending of A does not constitute a message, since the transmitter has exercised no choice in preparing A. Take the other electron, B, and apply to it, at the sender's choice, one of the four operations I, R_x^π, R_y^π, R_z^π; where I leaves it unchanged, R_x^π rotates it 180° about the x-axis, etc. Now send B to the receiver. The receiver is asked to select one spin component, y or z, and measure this same component for both electrons A & B. In either case the receiver recovers one bit of the two bit message encoded into B by the sender's choice of the operators {I, R_x, R_y, R_z}.

Figure 5.2: A page from Charlie Bennett's notebook, February 24, 1970 credit: Charlie Bennett

Connecting principles known from quantum mechanics into information theory was the catalyst that the theory of reversible computing needed. What physicists call the arrow of time, meaning its tendency to move in one direction, towards the future, is a function of entropy, or lost energy. At the quantum mechanical scale are a class of operations used to describe the time evolution of closed systems that are unitary, meaning they do not result in an increase in entropy. They preserve the state of the quantum system, and are therefore reversible. What Wiesner harnessed in his *gedankenexperiment* were examples of those unitary operations, and for Bennett, a connection from reversible computing to quantum mechanics was forged. Quantum information theory.

Slowly other unorthodox thinkers began pulling at this thread of quantum information, joining the efforts of Bennett and Fredkin. By the close of the decade, seminal work was being produced by pioneers such as Paul Benioff, Tommaso Toffoli, and Yuri Manin. Benioff, a researcher at Argonne National Lab, wrote a paper titled *The Computer as a Physical System*[11] in 1980 describing how a theoretical model of a computer could be thought of as a quantum mechanical system. Another paper in 1980 by Yuri Manin[12] similarly had talked about the theoretical possibility of quantum computing, though it contained scant details. Toffolli's *Reversible Computing* did not touch on quantum mechanics but it did introduce a new reversible logical gate, the Toffolli gate, advancing the understanding of the theory.[13]

Fredkin had also drawn his friend Richard Feynman to the topic of reversible computing. Feynman, a professor at Caltech in Pasadena, California since 1950, was widely renowned for his contributions to quantum mechanics, and by 1980 was a revered figure in the field. Through his books, public talks and his celebrated role in the investigation into the explosion of NASA's Challenger space shuttle, he would go on to become one of the best known physicists of all time, rivaling Einstein. His popularity was aided by his curious mind and his love for inquiry and oblique intellectual strategies above all else, and was famous for creative thinking.

Feynman's interest in computers traced back to his role on the Manhattan Project at Los Alamos, where some of the very first computations were performed by ENIAC for the design of the device. He'd led a team of physicists performing calculations the old-fashioned way, with brain power, devising a method for breaking down and running his teams' tasks in parallel, techniques that influence computer systems' architectures to this day. Fredkin had gotten Feynman interested in reversibility's potential to eliminate energy dissipation that posed engineering challenges when building computers, but he remained skeptical of Fredkin's enthusiasm for quantum information theory.

Finally Fredkin and Bennett collaborated to organize a conference on the subject, the previously mentioned Physics of Computation conference, jointly sponsored by MIT and IBM, in May 1981. Co-chaired by Fredkin and Rolf Landauer, the conference, held at an MIT property called Endicott House, would be an opportunity to get the little community of unorthodox thinkers together to thoroughly hash out the merits of quantum information theory. The group picture, seen in *Figure 5.3*, does not feature Bennett, because he was taking the picture, a role he'd repeat at similar gatherings of quantum information enthusiasts for years to come.

Over the weekend they presented their thoughts, debated various aspects of the idea of quantum information theory, and to close, Feynman delivered a keynote in the form of extemporaneous reflection on his perspective on the proceedings. In that talk, he expressed his belief that simulation systems from nature would be easier done with computers built on quantum phenomena like nature itself, and about the challenge of building such computers, he remarked "it looks like a darn hard problem to solve, by golly."[14] A certified genius calling a problem hard to solve is irresistible bait to a certain kind of mind, and the fledging field of quantum information science was now truly a legitimate topic to work on. The community formed at Endicott fanned out across the world to spread the gospel of quantum computing. The idea that information was a part of the physical world and that information theory should incorporate quantum mechanical ideas was increasingly taken as rote.

Figure 5.3: Group photo of participants in the Physics of Computation Conference, 1981. Photo credit: Charlie Bennett

Writing a review of the physics of computation published in 1982[15], Bennett used the ideas that began with Landauer to finally solve the mystery of Maxwell's demon. While measuring the energy of the molecules and even opening and closing the passageway between the chambers could be accomplished without expending energy, the demon was not violating the second law of thermodynamics. It could be assumed that the demon had finite storage for the information about the molecules' energy, and so at some point would have to erase some of that information to make room for new inputs. As Landauer had shown, permanent destruction is the indelible link between information and the physical world. Erasing a bit will always expend a minimum amount of energy, and so continues to increase the entropy of the system as a whole, as the second law dictates.

Bennett and the rest of IBM Research contributed far more than resolving this old *gedankenexperiment*. As the field spread across academic institutions, national labs and research institutions, there was no organization more invested than IBM Research. Landauer was already a larger-than-life figure in the company by 1981, and Bennett would end up being elevated as well to IBM Fellow in 1995[16]. Contributions, both theoretical and experimental, continued to be made by researchers at T.J. Watson and other IBM labs around the world, culminating in the public and commercial offerings starting in 2015 and reaching a kind of crescendo at the Think Q Summit in 2017.

The arc from Endicott in 1981 to T.J. Watson in 2017 would, for any classical technology, be the end of the origin story. The technologies we've seen emerge in the digital era would have transitioned from fundamental research questions to mere engineering challenges, finding ways to leverage the fresh scientific knowledge into ever more capable systems. Companies would be working to find a product-market fit with increasing investment from the private sector. The competitive struggle among the earliest stakeholders would play out, and IBM, or Google, or another company would emerge victorious, or power would end up being shared by an uneasy coalition. The ensuing years would see its maturation, increasingly productive output, unintended consequences and adaptive societal behaviors, as we have seen from the microprocessor, the Internet, mobile devices, and so many other technologies.

Quantum computing is not like all those familiar technologies. It was not unreasonable to ask if quantum computing even qualified as a technology in 2017, in the sense that no-one had definitive proof that a physical machine could be built that would be able to solve a practical problem better than classical computers. That proof remains elusive to this day. Quantum computing is so outside the bounds of what we are used to from digital technologies because of its reliance on the inherent weirdness of quantum physics, a field that still experiences its own controversies and skepticism.

References

1. Information is Quantum: What weird physical phenomena discovered a century ago have taught us about information and information processing, Charles H. Bennett, presented at Imaginado o Futuro Rio de Janeiro, 9 May 2018
2. `http://www.nasonline.org/publications/biographical-memoirs/memoir-pdfs/landauer-rolf.pdf`
3. Maddox, J. Maxwell's demon: Slamming the door. Nature 417, 903 (2002). `https://doi.org/10.1038/417903a`
4. Sophie Roux. The Emergence of the Notion of Thought Experiments. Thought Experiments in Historical and Methodological Contexts, Brill, pp.1-36, 2011. halshs-00807058

5. R. Landauer, "Irreversibility and Heat Generation in the Computing Process," in IBM Journal of Research and Development, vol. 5, no. 3, pp. 183-191, July 1961.
6. Page 9, http://www.nasonline.org/publications/biographical-memoirs/memoir-pdfs/landauer-rolf.pdf
7. https://philpapers.org/archive/ZUSRR.pdf
8. https://link.springer.com/article/10.1007/bf01857727
9. C. H. Bennett, "Logical Reversibility of Computation," IBM Journal of Research and Development, Vol. 17, 1973, pp. 525-532.
10. https://scottaaronson.blog/?p=5730
11. Benioff, P. The computer as a physical system: A microscopic quantum mechanical Hamiltonian model of computers as represented by Turing machines. J Stat Phys 22, 563–591 (1980). https://doi.org/10.1007/BF01011339
12. Yu. I. Manin, "Computable and Uncomputable," Advances in Mathematics 26, no. 2 (1980): 179-193.
13. Tommaso Toffoli, "Reversible computing," in Proceedings of the 7th Colloquium on Automata, Languages and Programming (Berlin: Springer, 1980), 632-644.
14. Richard P. Feynman, "Simulating Physics with Computers," International Journal of Theoretical Physics, 21(6/7), 467-488 (1982).
15. Bennett, C.H. The thermodynamics of computation—a review. Int J Theor Phys 21, 905–940 (1982). https://doi.org/10.1007/BF02084158
16. https://researcher.watson.ibm.com/researcher/view_person_subpage.php?id=10479

Subscribe to Deep Engineering

Join thousands of developers and architects who want to understand how software is changing, deepen their expertise, and build systems that last.

Deep Engineering is a weekly expert-led newsletter for experienced practitioners, featuring original analysis, technical interviews, and curated insights on architecture, system design, and modern programming practice.

Scan the QR or visit the link to subscribe for free.

`https://packt.link/deep-engineering-newsletter`

6

Universes of Information

In the development of ideas, primitive pre-ideas often lead continuously to modern scientific concepts... It is, for instance, possible to trace the development of the idea of an infectious disease from a primitive belief in demons, through the idea of a disease miasma, to the theory of the pathogenic agent.

— David Deutsch[1]

A theoretical foundation for computing in multidimensional spaces

By the time the motley crew assembled at Endicott House in 1981 to debate whether quantum mechanics might be useful for computation, classical computing was settling into a form that would remain familiar to us to this day. The mainframe and terminal were giving way to the personal computer, and the Osborne 1, the first portable computer, had just been launched, pointing towards our mobile computing present (admittedly, at 25lbs it wasn't *very* mobile)[2]. From the perspective in the mid-2020s, the computers and networks of the 80s seem feeble, with bandwidth, memory and storage measured in kilobytes, not gigabytes. However, that perspective is due to the extremely rapid rate of progress the technologies were enjoying. Thanks to Moore's law, the 68,000 transistors in Motorola's eponymous 68k chip[3] launched at the beginning of the 1980s was eclipsed by the Intel 80486's 1.5 million[4] by the end of the decade.

With that barreling engine of progress at full tilt, it seems peculiar that anyone would be looking for alternative approaches to computation, as were the Endicott House conference goers. For most, of course, the main motivator was pure curiosity — a very tricky puzzle to work out, regardless of the utility of any possible outcome. However, Feynman's explicit justification for the effort to define and establish quantum information science was to address the deficiencies of classical information science when it came to simulating natural systems. Classical computers, for all their increasing power, did have vulnerabilities which were already obvious to those who had been involved from the very start.

Feynman's central role at Endicott was not merely due to his stature in the field. At Los Alamos, his experience leading a team of human computers working on the math to model aspects of the Manhattan project put him at the very point of focus for the earliest digital computing projects at the University of Pennsylvania and Princeton's Institute for Advanced Studies. These machines were intended to perform the Manhattan project calculations faster than any team of humans, and, though neither was operational before the Trinity test shot, both were seen as huge leaps in our ability to simulate natural systems. Given his experiences with manual calculations and then classical computerized calculations, in conjunction with his continued work on quantum electrodynamics and other fundamental contributions to the understanding of nature, it was unsurprising that he was able to see the promise of using quantum computers to simulate quantum systems so readily.

The issue that Feynman and the others in those early days had seen with classical computers was the limitations of Boolean logic when attempting to simulate highly correlated systems, like those described by quantum mechanics. In these systems, each element is interacting strongly with many other elements in some way. A perfect simulation must represent those inter-relationships on a 1:1 basis, and so each additional correlated element adds exponentially to the connections necessary to represent it. Classical methods exist to work around this challenge. In an atom, for example, the electrons' spin interacts with that of all the other electrons present. If there are two electrons in orbit, you have two interactions to model. If you add an electron, you have four. Add another electron, eight. In addition, each electron interacts with the nucleus, and there are other interactions that will affect our model, all of which dramatically increase the computational burden of the simulation.

It's the classic parable of the grains on the squares of the chessboard which was first told in the 13th century about the game's creator in India[5] - when asked by the king what reward he wanted for his marvelous invention, he asked for one grain of wheat to be placed on the first square of the chessboard, two on the next square, four on the next and so on, doubling the number each time. The king laughingly agrees, saying it's a meagre reward, and then his advisors tell him that by the 64th square they will need to find a total of eighteen quintillion four hundred and forty-six quadrillion seven hundred and forty-four trillion seventy-three billion seven hundred and nine million five hundred and fifty-one thousand six hundred and fifteen grains of wheat. In other words, 1.1 trillion metric tons, or roughly 1,500 times the modern global annual production. That's exponential growth — two raised to the exponent representing the 64th doubling, or 2^{63}.

Nature, however, 'computes' the answer to these highly correlated natural systems so efficiently that we, as observers, are unable to perceive the computation time. An atom bomb just explodes without waiting for its force to be calculated by equation. A wooden table is an unimaginably vast collection of subatomic particles arrayed in such a way that it just sits there, being a wood table, in real time. The universe must possess some knowledge or mechanism that allows the efficient calculation of these highly correlated quantum mechanical systems, and so a computer built on quantum mechanical principles might afford those advantages as well. Most radically stated, information itself was being recognized as a natural phenomenon, intrinsic to the inner workings of the universe. A natural resource utilized by our abstract tools, instead of a man-made technology.

The origin of "information" as a distinct human technology is impossible to trace, given its self-referential nature - the tool is, after all, used to describe itself, first and foremost. We can imagine, however, groups of humans recognizing various observations as generally important to share, and repeat, and preserve, and at some point these observations become a rote category — knowledge, or wisdom, or lore. Over time, knowledge would have inevitably split into sub-categories. Optimal agricultural and animal husbandry practices for a region's climate, or the generally-accepted means for a community to make collective decisions, a favorite story told around the fire pit, or a recipe for a well-loved dish.

As communities grew into societies and specialization required a means of tracking the production and consumption of collective resources like grain, the first forms of mathematics and accounting emerged. Observations were recorded and preserved using marks in clay, on papyrus or other means. Accumulated observations and records represented the collected and preserved knowledge about a particular subject. Certain people began to devote their lives to understanding and extending the knowledge of particular subjects, and the information encoded in all those marks and pictures took on a life of its own, seemingly independent of its human origins.

Eventually Boole created his algebra, Turing created his abstract machine, Von Neumann built his machine, countless other women and men made their individual contributions and we had a universal logical machine equipped with a set of tools that could solve any problem expressed by the information theory understood in the middle of the twentieth century.

At its heart, computation as we understand it is classical information science transformed into practice through a set of gates or transformations that can manipulate the states of bits to create any Boolean logical operation. A NOT gate turns a 1 into 0 and vice versa. An AND gate takes two inputs, and if both are 1, the output is 1. If either is a 0, the output is zero. Other gates such as XOR ("exclusive or") and NAND ("not and") are also defined, and in a classical computer a set of these gates are implemented so that any logical "circuit" operation, or application of information science, can be carried out.

A classical algorithm such as a sorting routine or a Monte Carlo simulation is ultimately translated into a set of gates to be applied to a collection of bits, creating and executing a number of circuits which, in totality, carry out the desired action. The algorithm describes the logical steps required to get to the desired outcome — for example, given an unordered list of integers, compare the first with the second in the list, and if the second is smaller, swap their positions. Move to the second item in the list and the third, and repeat the comparison and swap if necessary. Continue down to the bottom of the list, then return to the first number in the list, and repeat. If you get to the end of the list and you haven't swapped the order of any two pairs, stop. You have performed a bubble sort, and the list now starts with the smallest integer at the beginning of the list and each item larger to the last, the largest number at the end of the list.

A bubble sort or any other algorithm can be translated into logical operations in a classical computer, a blazing fast sequence of circuits composed of gates dynamically constructed and executed until the final result is found. Every application we use classical computers for, from Minecraft to a simulation of a nuclear reaction, is composed of algorithms instantiated as circuits made of logical gate operations. An algorithm can be thought of as a practical expression of an information technology.

But, like an abacus, all of classical information is symbolic - a human abstraction meant to stand in for some actual thing. The beads on the rods of the abacus are merely representing the bushels of grain, though they can be used to perform arithmetic on them to forecast how far into the winter your stores will last. Bits are abacus beads instantiated in etched layers of semiconducting materials. They are a purely human invention, operating by rules of logic we defined, performing calculations in the way we invented. If there are limitations to classical information science, we must ask if those are a byproduct of the limitations of our own minds. Despite the efficacy of logic and mathematics, the question still remains, are they "real," or merely useful inventions of our evolved mammalian brains, the dreams of the universe, not the universe itself.

Quantum bits are realized through the manipulation of quantum mechanical systems that either exist in nature or are engineered to reflect a system from nature, and in that way are not merely symbolic representations of information that may be following our own arbitrary rules. Qubits contain systems that exist in nature managed and manipulated to achieve desired effects. Quantum information does come from the laws that nature is giving us, but just as for classical systems, human minds have had to divine and distill those laws into a workable technology. The same way we had to break down a pile of grains and figure out minimal universal laws to count them, we also had to learn how to break down piles of electrons and find out minimal laws to describe them universally, and that is what we call quantum logic.

Our own experience with quantum behaviors was necessary to lead us to quantum logic, but we validate that logic against the behavior of the quantum world, unbiased by our history of experiences, thus offering a new space for doing calculations. The universe is made of information, and quantum information science imagines a method for plugging into the computer that is the universe. Put another way, by exploring the physics of computation, Landauer, Bennet, Benioff, Manin, Fredkin, Feynman and the rest of the attendees of the conference in 1981 had realized that quantum mechanics could be used to reshape the technology of information science, to reform it out of the clay of the universe, and so surpass the limitations of the information science of our own invention. These tools would be instantiated from the dreams of the universe, dreams about itself.

After the Endicott conference, the focus of the work shifted from "why" to "how." In order to grasp the overwhelming otherness of a quantum computer, we're going to have to build more than a vague grasp on how they work, we're going to have to look at the lowest level building blocks. Don't worry too much, we won't be doing any real mathematics, just skimming the conceptual framework of how the whole thing holds together.

Building a quantum computer requires the implementation of operations that mimic the set of gates that power a classical computer, but instead of being implemented as variations in current in microscopic electrical components, these gates need to operate at the quantum scale. Thankfully the foundations of quantum mechanical theory defined in the early part of the twentieth century provided certain mathematical discoveries which lend themselves to logical operations. In particular, the work of a physicist named Wolfgang Pauli offered a promising starting point. In 1927 Pauli had published a paper that defined a set of matrix operations acting on the spin of certain types of quantum particles such as electrons, photons and protons. Pauli's matrices had been recognized as a powerful tool by Heisenberg, Born, and Pasqual as they had worked on the uncertainty principle, and over sixty years later they were seen as a useful tool again, this time in the development of a theory of quantum information.

These matrices, known now as Pauli-X, Pauli-Y, and Pauli-Z, describe "rotations" of a particle's "spin" around three axes. The use of scare quotes is entirely intentional, as the term "spin" is used conceptually only — in no way is a photon or electron spinning like a planet or top. The complete inadequacy of physical phenomenon or objects to act as metaphors for the quantum scale results in its reputation as being non-intuitive. Nothing we have ever experienced at our human scale, or even at the smallest and largest reaches we can observe with the help of optical or other sensory technological aides has any analog in quantum mechanics. Even the visible evidence of quantum phenomena, such as is exhibited by the double slit experiment, defies comprehension. Everything the physics describes at that scale is utterly alien, and yet it constitutes everything that exists.

In purely abstract terms, these spin-½ particles' state could be represented as a vector on a three-dimensional graph, and simplified further to a familiar x- and y-axis two-dimensional plot. The vector could be in two distinct states, with x equal to 0 and y equal to 1, or with x equaling 1 and y at 0, as illustrated in *Figure 6.1*.

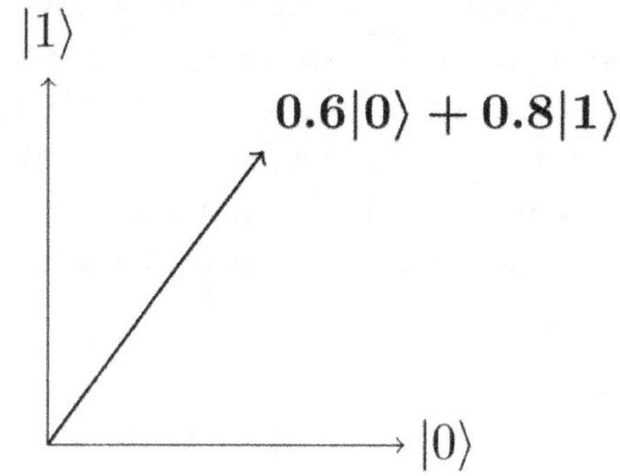

Figure 6.1: A two-dimensional plot of the vector representing a qubit's state

A vector can be represented as a matrix, with the two states of the particle represented as $|0\rangle = \binom{1}{0}$ and $|1\rangle = \binom{0}{1}$. Pauli's operations multiply the particle's state matrix against other matrices he formalized due to the transformations they carried out on the state. If a Pauli-X matrix multiplication is applied to a particle's state, the x and y values will flip to the opposite. If x is 0, it becomes 1, a 1 becomes 0, and the same happens to the y value. In the context of the conventions emerging for quantum computing in the 1980s, the Pauli-X was identified for its ability to act as a Boolean NOT operation.

You've noticed that the graph above doesn't use the large square bracket matrices to represent the computational basis states of a qubit, instead using $|0\rangle$ and $|1\rangle$ to represent the qubit's states. These symbols are named after Paul Dirac, a British physicist whose PhD thesis in 1926 was the first on the topic of quantum mechanics. Dirac notation is used in quantum information science to represent these vectors in the state space, using a symbol called "ket," or $|x\rangle$. In the 1/2 spin systems Pauli was working with, the [1,0] vector is designated as $|0\rangle$, while the [0,1] vector is $|1\rangle$. These two states are known as the "basis states." For computation, these map to our 0 and 1 states of classical bits. To represent a generic qubit of unknown state, we can write $|\Psi\rangle$. Schrodinger's Psi, used by physicists to denote an arbitrary quantum state.

This set of logical gates carried out by quantum mechanical operations may appear to merely represent classical computing with extra steps. However, with the classical basics out of the way we're ready for our first uniquely quantum operation — the Hadamard gate. Recall that the double slit experiment demonstrated that photons can be thought of as particles that act as waves under certain conditions, or waves that sometimes act as particles. Calculating the attributes of a particle behaving as a wave, according to Werner Heisenberg's uncertainty principle and the work of Erwin Schrödinger, the particle's state can be an indeterminate combination of possible states. This "superposition" is one of those quantum concepts that doesn't translate well into simple analogies because it has no near comparison in our scale. It is not the same as a flipping coin, or as being "all possible values at once," both of which are popular attempts to explain superposition which tend to make physicists very cranky. Nor is the cat actually alive and dead at the same time.

However, it is undeniable that superposition is a combination of probabilities of multiple states, and that when the state is measured, it "collapses" into a single state, with the probabilities of the alternatives irrevocably lost. Intuitively, this bizarre aspect of quantum information seems like it may be part of the mysterious "powers" the universe possesses as it computes its own existence so efficiently we can't observe it as it occurs. And so, we should have a gate operation for that.

Here a French mathematician named Jacques Hadamard provides the equations for our use. The Hadamard transform, first published in 1893, can be used to manipulate our vector so that its length remains 1, but its composition is now a combination of the $|0\rangle$ state and the $|1\rangle$ state with some probability. In terms of the vector depicted in *Figure 6.1* above, we have rotated it 90 degrees around the Y-axis followed by a 180-degree rotation around the X-axis. Mathematically, if we call the vector length of $|0\rangle$ alpha, or α, then the absolute value of α squared, or $|\alpha|^2$, is the probability of $|0\rangle$. Similarly, the probability of $|1\rangle$ is can be calculated as the absolutely value of its vector length squared, or $|\beta|^2$. Combined, the probabilities of α and β always equal 1, because, well, the probability of the universe existing is 100%, so far as we know. This can also be stated as the amplitude, or length of the state vector, remains 1, no matter what gate operations are applied.

While in superposition, the values of α and β are available for computation, but as soon as we measure the qubit, they collapse, are irretrievably lost, and the qubit is in one of its two basis states. The extra computational "space" during superposition is made even larger due to the complex nature of α and β — they are each represented by the equation a + b*i*, where a and b are real numbers, but b is the coefficient of the imaginary number *i*, where i^2 is equal to -1. You don't have to fully understand the math to recognize that $\alpha|0\rangle+\beta|1\rangle$ is *quite* a bit more sophisticated as a state space than that of a classical bit, that is, a 0, or a 1. If your algorithm requires elements of linear algebra like matrix multiplication or very high dimensionality, qubits got you covered. This concept of multi-dimensional space, with each qubit providing another dimension, maps to a concept called Hilbert space, named after David Hilbert, the same German mathematician who had posed the *Entscheidungsproblem* that had inspired Turing. Hilbert had hit upon the idea of infinite dimensional Euclidean spaces while working on differential and integral equations.

Crucial to quantum operators is the preservation of the total of the probabilities of the basis states summing to one. This total probability is represented by the length of the state vector of a qubit, which is also equal to one. Recall the fascination with reversible computation that helped motivate the investigation of quantum information by Charlie Bennet, Ed Fredkin and others. Many quantum gate operations such as the Pauli matrices or the Hadamard gate preserve the amplitude of the state vector as 1, and are said to be unitary, so that if the entries in the matrices are all real, the operations can be applied twice to a qubit, which will return to its original value, or "identity" matrix. In terms of the vector on a 2D plane, you can think of this as applying the same rotation twice, returning to the original value. The computations are reversible.

Additionally, unitary gates mean that the evolution of a closed quantum system, that is, one not being measured, is deterministic and therefore can be used for logical computation. Interestingly, the collapse of a superposition is considered probabilistic, with an uncertain outcome rather than deterministic. Despite the implications of this upsetting Einstein and others, it is another truly unique attribute of quantum computing, part of the native "language" of the universe, and a clue to the immense potential of quantum computing.

Even adding the Hadamard gate, our work so far can seem to be a mere projection of Boolean logic onto the quantum state space — a reconstruction of the "artificial" information science of classical computing. In order to support logical operations needed for anything humanly recognizable as an algorithm this set of building blocks is necessary. It does not, however, limit the complexity of the quantum state space. The two-dimensional projection is the simplest representation of a qubit. The other Paul matrices and another set of gates known as phase gates operate on other axes. Pauli-Y gates operate on the y axis and Pauli-Z and phase gates operate on the Z axis. Considering all three axes enables one to visualize a three-dimensional projection, represented by the Bloch sphere, as seen in *Figure 6.2.*

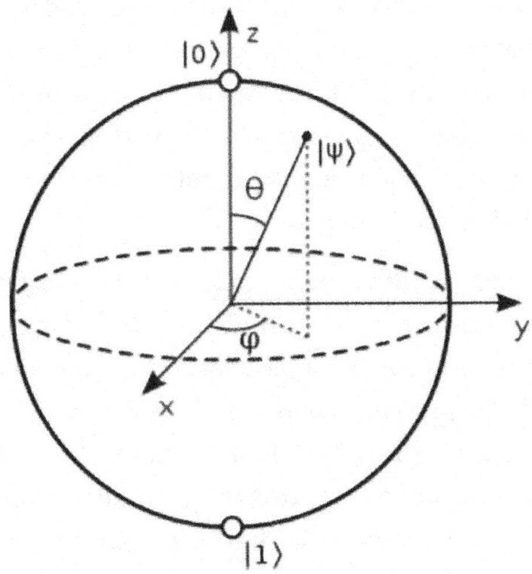

Figure 6.2: A Bloch sphere representation of a qubit state

This three-dimensional projection reflects full informational complexity of a single qubit in quantum computing. The 2-dimensional projection contains the classical 1 and 0 values as well as the superposition of those basis states, created by a Hadamard gate. Pauli-X, Y, and Z gates flip those basis states along the named axis. Phase shift, or S-gates, rotate the state around an axis. If this is starting to feel abstract, the key takeaway is this: where a classical bit is a switch with two positions, a qubit is a point on a sphere — it can encode vastly more information, and manipulating it requires a richer set of operations.

The collection of the gates we've examined so far are referred to as Clifford gates, and include the CNOT, or controlled NOT gate, which acts on two qubits to entangle their states. Clifford gates can be used to create a Hilbert space with a theoretical maximum number of dimensions limited by the number of qubits in the system. However, they do not represent capabilities unique to a quantum computer. In fact, Clifford gate operations are possible to simulate on classical computers with a relatively high efficiency, and therefore don't represent a path to a substantial performance advantage from hypothetical quantum computers.

A critical paper in the pre-Endicott work that helped crystallize the field was *Reversible Computing*[6], published in 1980 by Tommasso Toffoli. Toffoli is an Italian physicist and computer scientist who not only set the stage for Endicott, he attended the conference, attracted by the promise of unitary quantum operations' reversibility. In his paper Toffoli described a gate with three inputs, where if the first two bits are set to 1, the value of the third bit will flip. This gate's operations are reversible and can be used to create a universal set of logical gates, able to solve any computable problem.

Implementing Toffoli gates, also known as controlled-controlled NOT, or T-gates, in quantum information science achieves the same goal as in a classical system. Pairing Toffoli gates with the Clifford set of gates achieves universality, allowing a quantum system to solve any computable problem. What's more, its operations can perform tasks that are not efficiently simulated on a classical computer, and therefore it held the promise of significant quantum advantages to computational performance.

Drawing on work by a disparate set of mathematicians, theoretical physicists and computer scientists stretching out a century or more, the raw ingredients of a science of quantum information were gathered together. As a whole these gates were needed to fully reformulate information theory through the principles of quantum mechanics. In the decade after Endicott, building on the earlier work of Paul Benioff, Manin, Toffolli and others, a British physicist named David Deutsch worked towards this formalized model of universal quantum computing. Deutsch recalls a discussion with Charlie Bennett in the early eighties in which he complained that working on a theory of computational complexity seemed to him a waste of time since it would always be dependent on the capabilities of a given computer — there was no idealized form of a universal computer. Bennett offered his contrary view, that "'...the thing is, there is a fundamental computer. The fundamental computer is physics itself.'"[7] In 1985, Deutsch published *Quantum theory, the Church-Turing principle and the universal quantum computer*[8], which extended the prior work to a formal theory of quantum computing.

References

1. Deutsch, David. "On Wheeler's Notion of 'Law without Law' in Physics." Foundations of Physics 16, no. 6 (June 1,1986): 565–72
2. 'Osborne 1 Portable Computer | Science Museum Group Collection --- Collection.sciencemuseumgroup.org.uk', n.d. `https://collection.sciencemuseumgroup.org.uk/objects/co8094436/osborne-1-portable-computer-personal-computer.`
3. `https://en.wikipedia.org/wiki/Motorola_68000`
4. `https://en.wikipedia.org/wiki/I486`
5. Tahan, Malba (1993). The Man Who Counted: A Collection of Mathematical Adventures. New York: W.W. Norton & Co. pp. 113–115. ISBN 0393309347
6. Toffoli, T. (1980). Reversible computing. In: de Bakker, J., van Leeuwen, J. (eds) Automata, Languages and Programming. ICALP 1980. Lecture Notes in Computer Science, vol 85. Springer, Berlin, Heidelberg. `https://doi.org/10.1007/3-540-10003-2_104`

7. Rivka Galchen, "Dream Machine," The New Yorker, May 2, 2011, n.d. `https://www.newyorker.com/magazine/2011/05/02/dream-machine`.
8. Deutsch, D. (1985). "Quantum theory, the Church–Turing principle and the universal quantum computer" (PDF). Proceedings of the Royal Society. 400 (1818): 97–117. Bibcode:1985RSPSA.400...97D

Part 2

Reality

That quantum computing might be theoretically possible does not answer the question regarding whether or not they are worth building. The theoretical work of the 90s included some compelling proof that quantum computers might be able to solve certain problems that were impossible for classical computers. That one notable example threatened to undo all the encryption in the world caused a significant amount of interest in the esoteric field, and the race to try to build a machine was on. The path was by no means clear – in many ways the field was at a stage analogous to the computing industry before the invention of the transistor. And so, a hunt for the quantum equivalent of a transistor was on. Quantum bits, qubits, can be instantiated in many physical forms, all of which draw on the prior century of experimental quantum physics. We'll examine the roots of several of the earliest and most well-known of the qubit modalities, from physics experiment to industrial technology. Along the way we'll build a strong understanding of what is desirable in a qubit, why it is so hard to make them well, and how to get a sense of any particular qubit's chances to make the transition to commercial viability.

7

Shor, and Everything After

If the eternal dance of molecules

Is too entangled for us mortal fools

To follow, on what grounds should we complain?

Who promised us that Nature's arcane rules

Would make sense to a merely human brain?

— *This is No Clockwork Universe*

By Peter Shor[1]

Algorithms paint arrows towards future applications

Contemplating the years following Endicott, it can be difficult to understand what kept people engaged in the work in a field that had yet to prove if it would be good for anything. Deutsch's quantum update to the Church-Turing thesis, as important an achievement as it was, did not provide any justification for the effort it would take to to replicate the effort it had taken to build a Turing machine in the 1940s. Classical computing is universal, meaning that it is theoretically possible to solve any problem that can be properly expressed. Quantum computers, thanks to Deutsch's work, had a theoretical proof that they were capable of exactly the same feat, no more and no less. In order to justify the effort, it would try to build these theoretical devices, they were going to have to offer the promise of exceeding the capabilities of classical computers.

It seems as though this question must have been preoccupying David Deutsch as well, because in 1992, he and his collaborator, Richard Jozsa, proposed an algorithm that provided the first provable quantum advantage over a classical computer. Provided a function hidden in a hypothetical black box, the algorithm would determine if the function's output would be "constant" or "balanced." A constant output function would take *n* bits as input and return the same value no matter what the input was, and a balanced output would produce 1 for exactly half of the input values and zero for the other half of the inputs. That is, if your input is 4 bits of data, the constant function would always output 1 no matter what our input value was, while a balanced function would produce 1s for half of the possible values of the input, and 0s for the other half. The challenge is to properly identify the nature of the black box function in as few iterations as possible. Taking a classical approach, the worst-case solution would take just over half of the possible input values. In other words, take our 4 bits, which provide 16 possible input values. Suppose our black box function is balanced, and that it will output 0 for values 1-8, and 1 for values 9-16. If we start with our input set to 1, our output will be 0. If we increment the input by 1 and try again, we'll get another 0. At the 9th iteration, the black box will output 1, and we'll have proven it is balanced. If the black box was constant, the 9th iteration will also be 0, and its nature will be revealed. Our classical algorithm will, in the worst case, with our *n* of 4, take $2^{(n-1)}+1$, or 9 iterations to determine the nature of the algorithm, constant or balanced.

Deutsch-Jozsa uses quantum information to perform a nearly magical act - it invokes the black box function twice, and can determine whether it is constant or balanced without fail. It does so by encoding the input values into quantum bits and applying a quantum operation that puts them in superposition. It is then provided as input to a quantum version of the black box function, and after the second run, we have our answer. This was a demonstration of what Deutsch called "quantum parallelism," because it allowed for multiple input values to be tested simultaneously, reaching an answer very efficiently. How efficiently? The classical version of this algorithm required in the worst case $2^{(n-1)} + 1$ iterations to find an answer, where *n* is the number of bits provided as input. In other words, if our *n* is 4, as we saw above, the worst case will take 9 iterations. If our *n* is 8, we may need a maximum of $2^{(8-1)} + 1$ inputs to get our answer, or 129. In both cases, in every case, Deutsch-Jozsa does it in 2.

You may be wondering, what good is an algorithm that can perform this task? What is the point of this task at all? In fact, the point of this task is that it was designed to be particularly difficult for classical information theory and particularly yielding to quantum information theory. It has no other value than as an opportunity for the quantum theorists to show off. But what a result! What Deutsch-Jozsa proved was that a quantum algorithm could solve certain problems exponentially faster than classical approaches. As our value of *n* increases, the worst-case number of classical iterations increases exponentially, so that doubling the size of the input set to 16 bits increases our worst-case number of iterations to 32,769, and doubling again to 32 bits increases to over two billion! Deutsch-Jozsa still gets an answer in one go, provided we have enough quantum bits to represent our set of inputs. This was the first proof that a quantum computer could be not just faster, but exponentially faster — and that the gap would only widen as the problem grew.

This was the result that the theorists working in the field were waiting for — a verifiable speed up produced by a quantum algorithm — a justification for investing more time and effort into exploring this new frontier. Leading theorists from the University of California Berkeley, Ethan Bernstein and Umesh Vazirani, leveraged Deutsch-Jozsa to create their own algorithm, Bernstein-Vazirani, which, instead of determining whether a function was balanced or constant, tried to learn a string encoded in a function. Their goal was to provide a more generally applicable framework for evaluating the complexity of quantum algorithms.

It's worth pointing out that what these quantum algorithms did was prove there was an advantage for at least one kind of problem, but in no way did it suggest there would be an advantage across the board versus classical computers. This was an early confusion that drove some of the hype and skepticism around quantum computers, traces of which can be found to this day. No-one has ever suggested that quantum computers would be good at spreadsheets or video games or other algorithmic tasks which have driven the design and optimization of classical computers decade after decade. Instead, what excited information theorists was the possibility that quantum computers could carve out new territory in the field of computational complexity, and be able to solve certain types of problems that were too challenging for classical computers.

The study of computational complexity has its roots in the same work that spawned computers, and began to be formalized with the Church-Turing theorem and subsequent work to refine it. Essentially, in complexity theory there is an imaginary "problem space," referred to as PSPACE, which contains all possible problems. Within that space, as illustrated in *Figure 7.1*, there are different classes of problems characterized by the amount of time it takes to arrive as a solution via computation. The class of problems that can be solved in "polynomial time" are called P, and are considered to be solvable efficiently on classical computers. They are referred to as polynomial because the time to arrive at a solution scales linearly with the difficulty of the problem. In other words, a P problem that is solved in x time would, if you were to double its complexity, still be solvable, now in $2x$ time.

Other classes of problems are not as yielding, and are referred to as NP and NP-Complete for "non-deterministic polynomial" time scaling. While NP problems are solvable, no efficient algorithm is known for them, and in the worst case the time they take to solve may scale exponentially, so instead of $2x$ for a problem that is doubled in complexity as we saw with the P example above, an NP problem may take x^2 or even 2^x time. What the work of the first pioneers in quantum algorithms suggested was the existence of a new complexity class, a class that contained not just P, but some elements of the NP class. These NP problems that potentially scaled exponentially as their complexity increased may be solvable in polynomial time.

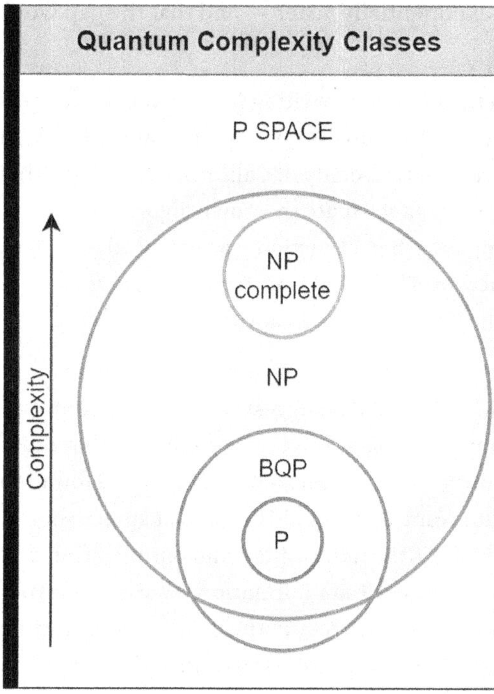

Polynomial Space Problem (PSPACE)	Space requirements are polynimial bound e.g. NxN Chess
Non Deterministic Polynomial Time (NP)	Solutions can be guessed and verified in polynomial time e.g. Graph Isomorphism
Non Deterministic Polynomial Time Complete (NP Complete)	Class of problem for which no efficient solution is available e.g. Travelling Sales Person
Bounded-error quantum polynomial time (BQP)	Problems can be efficienltly solved by quantum computer e.g. Factoring Discreet Log
Polynomial Time Solvalbe (P)	Efficiently solved by both quantum and classical computer e.g. Graph Connectivity

Figure 7.1: Computational complexity classes, including the quantum classes. Figure 7.1 Sahu, Himanshu & Gupta, Dr. (2023). Quantum Computing Toolkit From Nuts and Bolts to Sack of Tools.

This new quantum complexity class became known as bounded-error quantum polynomial time, or BQP. The excitement that Deutsch-Jozsa and Bernstein-Vazirani generated boiled down to the fact that quantum computers might be exponentially faster than classical computers for these tasks.

To excite interest in those whose research did not immerse them in computational complexity theory, however, it was going to take a proof with some practical purpose. The next step on that journey was taken just two years later, by Dan Simon, a researcher working at the University of Montreal. Simon's algorithm didn't do anything useful either, at least not directly. It did, however, demonstrate some very interesting attributes in terms of the task it tackled. Simon expanded on a concept that Bernstein-Vazirani had extracted from Deutsch-Jozsa, called a quantum oracle. This was an abstraction of Deutsch-Jozsa's black box, in this case Simon used it to speed up his solution to something called the "hidden subgroup problem," or HSP. Thanks to quantum parallelism, Simon was also able to achieve an exponential speed up, reducing the time from 2^n to simply *n* time. HSP is a technique that can be used to solve problems where you are trying to find an optimal path, called a graph problem, for example the "traveling salesman problem," where the challenge is to find the most efficient route that encompasses all the salesman's stops. More generally, HSP can be used to find the period, or hidden repeating structure, of a signal. This last fact proved highly suggestive to another researcher, Peter Shor.

Shor was a researcher at AT&T Bell Labs, where, after seeing a lecture from a visiting Charlie Bennett, developed an interest in quantum information science. He was drawn to Simon's algorithm in particular, realizing that it could be adapted to calculating discrete logs very efficiently. Discrete logarithms, or logs, are the exponent required to apply to one number to get another, in other words given real numbers a and b, the $\log_b(a)$ equals x such that $b^x = a$. This is easier to demonstrate than to explain. In base 10, the powers of 10 are 0.01, 0.1, 1, 10, 100, etc. For any number in this list, you can easily compute the log, for example, the $\log_{10}(100)$ is 2, and the $\log_{10}(0.01)$ is -2. In other words, 100 is 10 to the power of 2, and 0.01 is 10 to the power of -2. This example for base 10 is simple, however in classical computing calculating discrete logs of other integers is considered intractable — not in the sense that it is unsolvable, but that producing a solution will take so long as to not be useful. Shor's adaptation of Simon's algorithm, on the other hand, yielded an exponential speed up in its solution, leveraging the oracle, quantum parallelism, and a variation of Simon's HSP solution — put simply, it could produce an answer in a practical amount of time.

This initial breakthrough was sufficiently impressive that Shor was invited to give a talk at Bell Labs by Len Adelman, a very highly regarded mathematician and cryptographer. Adelman introduced Shor by saying to the effect that, "there are two kinds of geniuses. There are the DaVincis, whose inventions are beautiful but ultimately don't work, and the Babbage's, whose inventions are beautiful and eventually do work. I don't know if Peter is a DaVinci or a Babbage." This compliment made such an impression on Shor that it is a permanent part of his memory of the extraordinary year he had in 1994.

Members of the audience must have been impressed as well, because the news of Shor's algorithm made its way to Umesh Vazirani at UC Berkeley seemingly overnight. Vazirani actually recognized something even more powerful hidden in Peter's initial algorithm, and called him on the weekend to ask if he'd looked at factoring. In fact, Peter had. As he tells the story, after the talk on a Tuesday, he realized his work could be adapted to efficient factoring, which he was able to describe to Vazirani over the phone just a few days later, despite suffering from a cold at the time.

In April 1994 Peter published his work, and his factoring algorithm, what would be known as Shor's algorithm, hit like a lightning bolt. Prime numbers are those which, as factors, have only themselves and 1 — no other number divides evenly into a prime. It is, of course, trivial to multiply two large prime numbers, *n* and *p*, together, to arrive at their product, *q*. However, it is very inefficient, given *q*, to reverse engineer the prime factors of *q*, namely, *n* and *p*. The classical algorithms known for this task have to "brute force" a solution by trying to divide any candidate answer by all numbers smaller than it to prove its prime status. The larger the potential answer you are testing, the more numbers you need to test it with, and the larger the number you're trying to factor, of course, the more numbers might be one of its prime factors. The time taken to solve this classically rapidly expands into intractable territory — while Shor's algorithm can find those primes exponentially faster.

Shor accomplished his magic with a very original approach to the problem of factoring and unique quantum mechanical phenomena. First, inspired by Simon's oracle, the algorithm prepares a state whose projection onto each of the system's basis states, our measured, probabilistic combination of |0> and |1> from before. Each of these basis states represents a possible answer, and collectively they represent the set of all possible answers. Then, he very cleverly recast the factoring problem as one of finding periodicity. Using a classical algorithm to find the period at which the factors of the input number repeat, Shor then uses that period value to drive a quantum version of a function called a Fourier transform across the quantum bits in superposition, utilizing what Deutsch called quantum parallelism to achieve his exponential speed up to produce all the number's factors. A simple classical post-process tests the factors to find the primes. The classical pre- and post-processing added some overhead to the quantum computation, but not enough to dull its impressive performance advantage.

The real impact of Shor's algorithm wasn't just that it did something much more efficiently than classical computing, it had to do with what exactly it was doing. Cryptography is a set of techniques for obscuring information one wishes to keep private, and in the digital age we have relied on one technique in particular, namely, using the prime factors of very large numbers as our keys to unlock the secret information. The classical factoring problem is of sufficient complexity that it forms the basis for many modern cryptographic systems, most notably Rivest-Shamir-Adleman[1] (RSA) — yes, the same Adelman who hosted Shor for his first public talk on his discrete log algorithm. RSA and other cryptographic algorithms like Diffie-Hellman, are public-key or asymmetric key encryption schemes. These cryptographic keys rely on the "one way" nature of prime factoring, making encryption very efficient for anyone possessing the public key but highly inefficient without the private key. The private key contained knowledge of the primes required to decrypt the message. Modern cryptography on classical computers relies on the fact that there are no shortcuts to factoring a large number to its primes, and Shor seemed to have just invented such a shortcut.

And so, overnight quantum computers went from an odd thought experiment that might not do anything for anyone, to seeming to bring the movie *Sneakers* to life. In the film, released in

1992, a technology is created that can decrypt all secret data, instantaneously. Shor's algorithm would not deliver instantaneous results, but it could reduce the code breaking time drastically. Some models estimate a 2048 bit RSA code, which would take classical computers trillions of years to break, would yield to a quantum computer with about 4,000 error-corrected logical qubits in about 10 seconds[2]— which, with current error correction schemes, translates to millions of physical qubits, a scale no one has come close to achieving. In another cosmic coincidence, when the film makers wanted to make the script for *Sneakers* seem more believable, they hired a mathematician with significant cryptographic experience to act as a consultant — Len Adelman.

Shor's algorithm was no Hollywood fantasy, and the intelligence agencies of the world sat up and took notice. Quantum computing was finally graduating from an eccentric's hobby to a legitimate technology with stupendous potential power. The catch was, of course, that qualifier on the estimate that it would take a computer with 4000 perfect qubits. In 1994 there were no systems with that description, and there was no well-defined path to get there. All this work, up to the time of Shor's publication of the factoring algorithm, had been theoretical. No operating qubit had been created, and no one knew if it was even possible to build an operational quantum computer. Attention shifted from the theory to potential realization. Given the sensitive information that Shor's algorithm seemed to threaten to unlock, the field found itself suddenly very well-funded, and a first quantum bubble began, fueled by public sources of funding.

References

1. `https://math.mit.edu/~shor/No_Clockwork_Universe.html`
2. R. L. Rivest, A. Shamir, and L. Adleman, "A Method for Obtaining Digital Signatures and Public-Key Cryptosystems," Communications of the ACM 21, no. 2 (1978): 120-126.
3. `https://postquantum.com/post-quantum/4099-qubits-rsa/`

8

Experimentalists Just Wanna Have Fun

> *"This scheme, like all other schemes for quantum computation, relies on speculative technology, does not in its current form take into account all possible sources of noise, unreliability and manufacturing error, and probably will not work."*
>
> *— Rolf Landauer*

Just how does one build a quantum computer?

The foundations of quantum computing were built by theorists' equations, from Wiesner, Deutsch, Shor and the rest. Now that it had been established that universal quantum computing was theoretically possible; and that it could do certain things *much* faster than classical computers, attention began to shift to the experimentalists working in their labs with lasers, liquid helium, and other hazardous materials. The fact that Shor had attracted the attention of certain government agencies, who now wanted to know who could build a machine to run the algorithm, and how much it would cost, was only to their advantage. The public funding sluices opened up, and anyone with a reasonable idea about how one might build a qubit found themselves flush with budget to pay postdocs and buy all kinds of equipment and components.

The theorists had begun working towards realizing physical quantum bits by describing them in the abstract, using quantum mechanical terms. All the experimentalists needed to do was to identify quantum systems that could fulfill those characteristics and away they'd go. Specifically, the system had to have two or more distinct states measurable in attributes such as spin, polarization, or energy level that could be reliably be induced, maintained, manipulated and read out. These two-level systems had a nice new portmanteau, "qubit," for quantum bit, coined by Ben Schumacher in 1993 in a paper that was published in 1996[1]. With a slick new name and plenty of funding, inventing qubits ought to be easy.

In Wiesner's paper, he'd used the polarization of photons as the attribute used to store and represent information, and in 1989, Charlie Bennett and John Smolin at IBM Research had created a working quantum communication demonstration using polarized photons supplied by lasers, seen in *Figure 8.1*. Peter Shor had, in fact, first encountered the field via a talk that Bennett gave at Bell Labs about that experiment.[2] Photons were attractive because polarization is easy to measure and easy to maintain over a long period of time. They are stable at room temperature, so far as quantum systems go, and not sensitive to disturbances from their surrounding environment, or noise. However, the reason photons are not susceptible to noise is that they are very resistant to interactions in general, and are therefore difficult to manipulate. Operations critical to performing quantum computation, like entanglement, can only be accomplished probabilistically. In other words, the experimentalist can set up the conditions which may induce entanglement between photons, but the particles may or may not cooperate. This made progress with photonic systems slow after the Bennett-Smolin experiment.

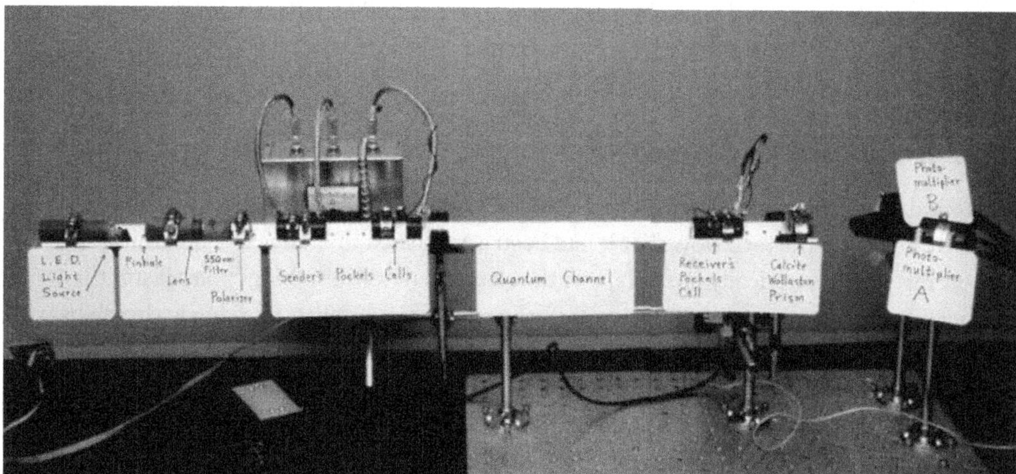

Figure 8.1: Bennett and Smolin's experimental implementation of BB84 protocol for quantum communication. Credit: Charlie Bennett

Following the Endicott conference, thankfully, experimental physicists had been evaluating the various themes of their research in collaboration with theorists, looking for potential qubit candidates. Early proposals to create novel systems tended to be more hypothetical than practical using techniques that didn't exist yet combined in fairly far-fetched schemes which, if you squinted your eyes and tried very hard to believe, *might* function as computers. The first example of this occurred in 1988, when two Japanese researchers, Yoshihisa Yamamoto and K.

Igeta, proposed using atoms for two qubit operations (e.g. entanglement) and photons to transmit the qubit states around the system. On paper the scheme suggested it would support a universal set of quantum operations including the CNOT gate that produces entanglement, but implementation details were thin.

A set of formal criteria for suitable qubit implementations began to crystallize thanks to the efforts of another IBM Researcher, David DiVincenzo. Informally circulating in the community trying to conjure up experimental ideas that could be useful, he ended up publishing them in 2000, by which point they were as follows:

1. The ability to scale physically to many qubits, with the qubits uniform and with predictable behavior — in plain terms, you need to be able to make a lot of them, and they all need to behave the same way.
2. The ability to initialize the state of the qubits to a well-defined starting point, such as | 000...> — you need a reliable reset button.
3. A set of quantum operations, or gates, that could support universal logical computation — the system needs a complete instruction set.
4. The ability to remain coherent for much longer than it took to apply a gate — in other words, low noise and high stability. The qubits need to hold their quantum state long enough for you to actually do something with them.
5. The ability to measure the state of the qubits accurately — when you're done computing, you need to be able to read the answer.[3]

The DiVincenzo Criteria were emblematic of an important shift among the physics community towards thinking in terms of systems engineering. That is, the criteria were about the characteristics of the device being designed rather than the typical focus on a stand-alone experiment. Physics labs tend to produce instruments that, with months, years, sometimes even decades of preparation, can perform a set of operations that yield data for the scientists to analysis *ex post facto*, a process that may again take months or years to perform. The resiliency, reliability, manageability and other aspects of the instrument's behavior as a system are typically not priorities.

For example, the first criteria addresses scalability, or the ability to make not just one or two qubits but many orders of magnitude more. Classical computers, after all, had only been made possible by the ability of silicon microprocessors to scale into the hundreds of millions of transistors. Initial calculations of the number of qubits required to factor a number on the scale of those used in contemporary cryptography using Shor's algorithm estimated that it could take tens of millions of qubits. Any serious attempt to build qubits would need to address that requirement.

The criteria that the qubits would need to implement a universal set of gates represents the overlapping of the theorists' world with that of the experimentalists. A Turing-complete quantum computer, capable of computing all that is computable, would require physical devices that could carry out all the linear algebra needed to implement the required logical operations. In other words, Deutsch and Shor may have been able to conjure the mathematics needed to reimplement a Turing machine using quantum mechanics, but only experimentalists could design and build machines that would bring that to reality.

The remaining three criteria, that the qubits must be able to be initialized into a well-defined initial state, to withstand noise and retain coherence longer than it takes to carry out operations, and to have their resulting data read out reliably, all speak to the stable operation of a hypothetical machine. Here was the central challenge faced by those seeking to build these devices: engineering a stable system that exposed and made stable the behaviors of the sub-atomic realm was far more challenging than anything ever accomplished to date. The scale of quantum mechanics defies our sense of scale, with dimensions starting at billionths of a meter, or nanometer, in numerical terms 0.000000001 meters, 10^{-9}, and descending from there. Classical microprocessors in the late 90s were being fabricated with circuits at the scale of about 150 to 200 nanometers.

The seeming impossibility formed part of an irresistible draw for ambitious experimentalists, and the late nineties saw a burst of creative productivity. A number of more feasible proposals were floated by researchers in Europe, the Americas and Asia, ideas cross-pollinating around the world. Various other domains within the experimental community that had produced exotic lab instruments began to be re-evaluated in the light of the requirements laid out by DiVincenzo. In fact, the concepts for qubit modalities that would become dominant in the 2010s can all trace their roots to the period between Shor's algorithm in 1995 and the end of the decade. Qubit designs based on cutting edge experimental physics devices such as ion traps, neutral atom arrays, superconducting circuits, even more speculative ideas like quantum dots and electron spins were first defined in papers from this era, though truly successful implementations would take another decade to be achieved.

At a high level, the explorations began to fall into two camps: those seeking to design systems that brought quantum phenomena up to a larger scale than was typical in nature, and those hoping to extend our engineered control down to the "native" quantum scale. One lab experiment seeking to manipulate systems at a quantum scale that was relatively mature by the 1990s was nuclear magnetic resonance, or NMR. The "nuclear" in NMR refers to the nucleus of an atom, which, when bathed in a strong magnetic field and then pulsed with a weaker radio frequency signal, produces an electromagnetic pulse of its own, characteristic to its own composition. In other words, electromagnetic signals can be used to coax an atom into revealing the subatomic, or quantum, state of its nucleus. This effect was first observed by

Isidor Rabi in his lab at Columbia University, where his experimentation established an enormous amount of knowledge about the atom.

Rabi was an incredibly gifted scientist, born in the Austro-Hungarian empire in 1898 and raised as an immigrant in the Lower East Side of New York City, he was an experimentalist *par excellence.* At Columbia University, where he took a position in 1929, he set about extending the result of a prior experiment carried out by Otto Stern and Walther Gerlach in Frankfurt in 1922. The Stern-Gerlach experiment stands as one of the first great validations of quantum mechanical theory. The Stern-Gerlach setup had been an adaptation and extension of Young's original double slit experiment from 1801, but instead of shining a light at gaps in an opaque screen, they aimed a beam of silver atoms through a magnetic field that could be varied in strength and direction. The atoms split into discrete spots on the detection area, forming an interference pattern similar to the photons in Young's experiment. This established that the magnetic "moment" of the atoms existed in discrete quantities, and therefore their angular momentum as they travelled from the source to the destination was also "quantized."[4]

Rabi's extension of Stern-Gerlach maintained the beam of atoms or molecules, achieved by heating a substance in the experiment's oven, and the variably tuned magnetic field, used to orient the resulting atoms' or molecules' magnetic moment in a specific way. After the initial magnetic field, the was subjected to an oscillating magnetic field that was perpendicular to the direction of the beam, and whose frequency could also be varied. When the frequency of the oscillating field matched the natural frequency of the magnetic moments of the particles, resonance would occur, which would cause a change in the orientation of the particles' magnetic moment. Finally, the beam hit a detector which would measure the intensity of the beam and reveal the magnetic properties of the atoms.

The resulting technique was called molecular beam magnetic resonance, and with this experiment, Rabi had invented the field that would be known as NMR, earning a Nobel prize for his work in 1944. Rabi's initial experimental results were further built upon by Felix Bloch and Edward Mills Purcell to use on a variety of liquid and solid material in 1946, work which earned them the Nobel Prize in 1952, and the basis for modern NMR devices was fully realized.

In the decades since its discovery, physicists had found many applications of NMR, using its ability to probe the inner state of materials for analytical chemistry and biochemistry experiments, a technique called spectroscopy, eventually leading to the commercialization of the technology as "nuclear magnetic resonance imaging" in the healthcare industry. The term "nuclear" evoked destructive weapons in the public imagination, which was avoided by shortening the name to magnetic resonance imaging, or MRI. Industrial development of MRI devices in the 1970s and 1980s meant that mature, stable, and relatively inexpensive devices were available "off the shelf" for physicists to experiment with, and so it was no surprise that it was considered as a platform for creating qubits.

As early as 1993 just such a proposal was made by the theorist Seth Lloyd of MIT, followed by an expanded proposal by David DiVincenzo in 1994. Missing crucial implementation details, the proposals languished for a few years, until 1997, when two separate experimental groups, one at MIT and the other at Stanford and UC Santa Barbara, published papers nearly simultaneously that provided the missing details, namely, how to implement gates using NMR devices.

Following the ideas laid out by the papers, a group at Oxford led by Jonathan A. Jones and Michele Mosca created two working NMR qubits and used quantum gates to run the Deutsch-Jozsa algorithm on them. The threshold between theory and experimental proof had been crossed, and history had been made, and, as if to further underline the event for posterity, a group of researchers from UC Berkeley, Stanford and MIT replicated their experiment, and so did Isaac L. Chuang, who had by then moved to IBM Research's Almaden Lab, seen in the lab in *Figure 8.2*.

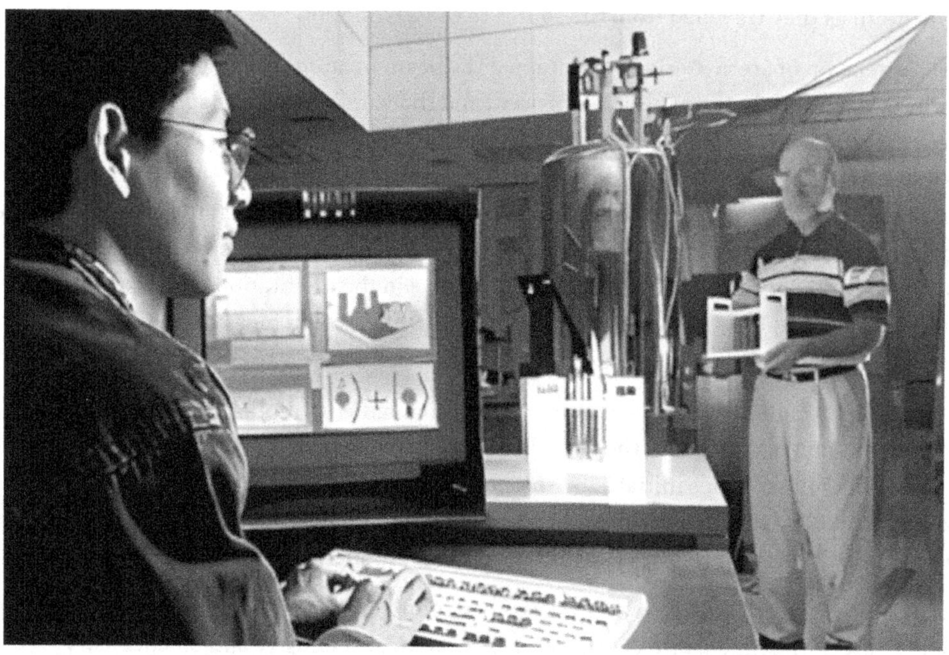

Figure 8.2: Isaac Chuang in the IBM Almaden lab working on computation using NME qubits.

These NMR qubits were not physical devices or even singular entities — they were created out of collections of molecules, called ensembles, that were manipulated with the magnetic and radio frequency fields in the NMR machine. The way the researchers implemented qubits in NMR devices was not dissimilar to the way the technology is used to scan our bodies. The strong magnetic field causes the protons in the hydrogen atoms to align, then the radio signals pulse and cause the protons to become unaligned, and when they stop the sensors detect the energy released by the protons reassuming their alignment. The different types of tissue in our bodies emit energy differently during realignment, which can be transformed into imagery. In NMR qubits, the basic NMR function could read out the spin of the molecules in bulk, and additional, custom tailored radio signals were used to modify their spin, implementing a universal set of logical gates. Different atoms within a molecule could be addressed separately, realizing multiple qubits out of the components of a molecule. It took at least 10^8 molecules to generate a strong enough signal to detect the state of the individual qubits. Control of these ensembles was, right from the beginning, the main challenge facing the NMR qubits.

Still, the experimentalists forged on, improving their techniques of control and measurement, and increasing the number of qubits from 2 to 3, and moving from Deutsch-Jozsa to a more complex search algorithm named after a researcher named Lov Grover. In addition to the challenge of controlling such massive numbers of molecules in a liquid all in perfect resonance, the NMR work faced theoretical challenges as well. A researcher named Samuel L. Braunstein published work with his colleagues that proved that none of the experiments on NMR qubits to date had included any quantum entanglement. Without entanglement it was very unlikely that a quantum computer made up of NMR qubits would ever out-perform a classical computer, since everything but quantum behaviors could be efficiently simulated.

Research continued, especially at IBM Research's Almaden Lab, where Isaac Chuang and colleagues continued to scale up their control systems, taking on ever-more ambitious algorithmic execution. In 2000, Chuang and team ran the first part of Shor's algorithm, and saw a path to be able to run the entire thing, they just needed seven qubits. This was a real test of the scalability of NMR qubits, because Chuang and his postdoc Lieven Vandensyden had to design and synthesize a molecule that would be stable and have seven atoms suitable to act as qubits. The candidate molecule was hard enough to specify, the atoms had to be sufficiently separated in their energy states for the NMR signals to act on them each in isolation. The group did find a suitable molecule, and then work began on synthesizing it. A full year later the sample was ready in a glass tube that could be fitted into the NMR machine. They prepared the state of the seven qubit system, building on work done at Los Alamos National Laboratory the prior year. The group then managed to run a full implementation of Shor, factoring 15 to its primes, 3 and 5[5]. Chuang moved back to MIT shortly thereafter, and, sensing that further advances in scale would be nearly impossible to realize, abandoned NMR qubits for good.

It would be easy to say NMR was a dead end, the Daguerrotype of qubits, abandoned as soon as more effective methods came along, but in fact the effort was quite valuable. First and foremost, successfully running Shor's algorithm was a moment that immediately earned a spot in the history of the field. Sure, it was only 15, but that record stood for a decade, finally being defeated in 2012, when another team using photons factored... 21[6]. More importantly it proved that it was possible, striking a blow against the skeptics and naysayers.

Even the criticisms leveled against the specific modality of qubits had value. What Braunstein's work pointed out was that it was not sufficient to make a computer out of a quantum system, that system needed to support truly quantum phenomena, and use them for the purposes of computation. The tricks employed by Deutsch, Jozsa, Simon, Shor and the rest of the earliest pioneers of quantum algorithms were leaning heavily on superposition and entanglement to achieve their speed ups, and without them all you were doing was building an extra complicated, less capable classical system. The need to verify the use of quantum entanglement in its operations has become a standard part of the evaluation of new devices.

Perhaps most importantly, the work on achieving control of NMR qubits impacted the development of control technologies for other qubits. The equipment used was being built for industrial users in a healthcare setting, and as such it had a robust embedded programming language to control the various parameters of the device. The quantum researchers leveraged that control language and built a sophisticated suite of software tools for initializing qubits, applying logical operations and measuring qubits states once calculations were complete, and these tools were shared and maintained by the various groups using similar hardware for their experiments. This was not the usual scenario for experimental physics, where if software was written it was typically hacked together for one researcher's very specific purpose, not documented, not usable by anyone else, and abandoned when the experiment was complete. This is, in fact, still very often the case in modern physics labs.

However, when the brief reign of NMR qubits ended, at least one notable researcher, Isaac Chuang, pivoted to an alternative qubit modality, and his postdocs and students started looking at what components of the NMR software toolkit could come along with them. That new modality being set up in Chuang's lab was qubits implemented using isolated charged atoms suspended in electromagnetic fields — known as trapped ions.

References

1. `https://journals.aps.org/pra/abstract/10.1103/PhysRevA.51.2738`
2. `https://ep-news.web.cern.ch/content/interview-peter-shor`
3. DiVincenzo, David P. (2000). "The Physical Implementation of Quantum Computation". Fortschritte der Physik. 48 (9–11): 771–783. arXiv:quant-ph/0002077. Bibcode:2000ForPh..48..771D. doi:10.1002/1521-3978(200009)48:9/11<771::aid-prop771>3.0.co;2-e. ISSN 0015-8208.
4. `https://pubs.aip.org/physicstoday/article/56/12/53/632269/Stern-and-Gerlach-How-a-Bad-Cigar-Helped-Reorient`

5. Vandersypen, L., Steffen, M., Breyta, G. et al. Experimental realization of Shor's quantum factoring algorithm using nuclear magnetic resonance. Nature 414, 883–887 (2001). `https://doi.org/10.1038/414883a`
6. Martín-López, E., Laing, A., Lawson, T. et al. Experimental realization of Shor's quantum factoring algorithm using qubit recycling. Nature Photon 6, 773–776 (2012). `https://doi.org/10.1038/nphoton.2012.259`

Get this book's PDF version and more

Scan the QR code (or go to `packtpub.com/unlock`). Search for this book by name, confirm the edition, and then follow the steps on the page.

UNLOCK NOW

Note: Keep your invoice handy. Purchases made directly from Packt don't require an invoice.

9

Nature's Perfect Qubits

It's easy for quantum computing to sound like a fanciful theorist's dream...But just extrapolating Moore's Law... somewhere around the year 2020 or 2030 is when we hit one bit per atom... So if we really want to keep getting faster...Quantum weirdness is essentially the only resource we have that's still untapped for computing.

— Neil Gershenfeld [1]

As Feynman said, "there's plenty of room at the bottom."

The argument for using individual atoms for qubits is simple - every atom is identical, and is perfectly consistent and predictable according to the rules of the physical world. This is quite unlike anything humans design and fabricate from materials with defects and minute variations. While NMR qubits used molecules as the basis for computation, others sought means to dive deeper into the quantum scale, isolating and manipulating individual atoms in systems that could support computation. In fact the story of physics is in large part the efforts of experimentalists to extend our perception and awareness down to the sub-atomic scale, a challenge that is difficult to imagine for those not in the field. The methods employed by these scientists to observe what is beyond the limits of our direct sensory observation are, by necessity, wildly creative.

Sub-atomic experimentation began in earnest in 1897, when Joseph John Thomson, a British physicist and Cavendish Professor of Physics at Cambridge University[2] conducted a number of experiments that established the existence of the electron. Thomson's experiments used a cathode ray tube, a device that consisted of a glass chamber where a vacuum was formed and electrical current was applied to each end. Applying a current, a beam would flow from the cathode, or negative electrode, to the anode, or positive electrode. Devices of this type had been used by scientists for generations in their experiments.

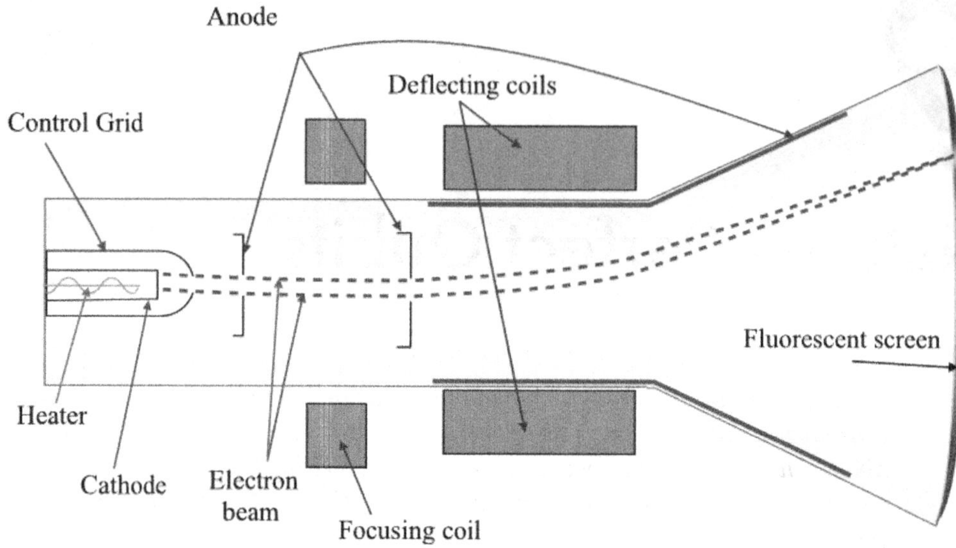

Figure 9.1: Cathode ray tube similar to that used by J.J. Thomson to experimentally establish the existence of the electron. By Theresa Knott - en:Image:Cathode ray Tube.PNG, CC BY-SA 3.0, https://commons.wikimedia.org/w/index.php?curid=100143

If this instrument looks familiar, it may be because you watched television between 1927 when Philo Farnsworth used a CRT designed in 1897 by Ferdinand Braun to create the first prototype and 2015 when the last television using a CRT was built. Before television, CRTs were used in oscilloscope and radar displays, and as we saw in chapter 4, were used as rudimentary digital memory devices by Von Neumann and his team at Princeton.

Before Braun came along, the original tube design had been modified by William Crookes in 1890 with the capability to produce negative and positive electrical charges midway along the tube, and these charges would deflect the ray from its normally straight path. Thomson was the first to measure the amount of the deflection compared to the amount of current. The amount of deflection from a straight path compared to the strength of the electrical charge provided a measure of the mass of the particles in the stream. Before Thomson's experiments, the atom was thought to be the smallest unit of matter and was believed to be indivisible and indestructible, a concept that dated back to ancient Greek philosophers like Democritus. In the early 19th century, the work of John Dalton and others formalized an atomic theory that described atoms as solid, indivisible spheres that made up the elements. This view was not without controversy and questions remain to this day about Dalton's methods and whose work he may have used without attribution, but by Thomson's time the model was generally accepted in the scientific community. According to this model, the smallest element, hydrogen, was indivisible and therefore the smallest mass in existence. However, the deflections produced by Thomson's experiments suggested that the ray was made up of particles thousands of times lighter than the smallest element.

The design of Thomson's experiment, and its implications for atomic theory, provides insight into the methods used to extend our understanding of the world beyond the reach of our senses. Unable to directly observe what was traveling through the tube, Thomson found a way to manipulate where the ray hit the fluorescing material on one end the beam by varying the charge affecting the beam's path. Having proved the beam was composed of something much smaller than any atom, Thomson named these particles "corpuscles," borrowed in part from Newton's corpuscular theory of light, and proposed that atoms were like a pudding with the negatively charged corpuscles embedded in them[3]. His work was tied back to experiments by George Johnstone Stony who, in 1891 coined the term electron for what he believed was the fundamental unit of electricity. This was the first experimental evidence of a subatomic particle, significantly altering the understanding of atomic structure, which previously was thought to be indivisible.

This experiment of Thomson's was not only groundbreaking because of the implied existence of electrons as a smaller component of the atom, but also because it paved a path toward mass spectrometry, and eventually, ion traps. By the early 1900s, Thomson had built upon his cathode ray experiments, modifying the instrument to bombard gases inside the vacuum tube with electrons and then discharge the electrons via the anode, leaving behind positively charged atoms of the gas. These positive ions were then attracted towards the cathode, and formed the basis of Thomson's analysis. Turning again to the magnetic and electric fields from the former experiment, Thomson could deflect these ions onto a detection screen. By analyzing the deflection patterns, he could infer properties like the charge-to-mass ratio of the ions.

Following the initial experiment, many advancements in technique and equipment were made by various experimentalists, becoming a common technique for identifying and quantifying molecules, based on their mass, referred to as "mass spectrometry." By the 1950s, however, limitations in the precision of measurements produced by the best experimental set ups inspired enhancements that would bring these ion beam devices closer to their eventual intersection with the field of quantum computing.

Wolfgang Paul, a German physicist not to be confused with Wolfgang Pauli, had created a non-magnetic mass filter in an attempt to achieve greater precision than was possible with the magnetic deflection methods of spectrometry to that time. The device was developed with his student, Helmut Steinwedel, in 1953, and consisted of four parallel metal rods, called a quadropole. As seen in *Figure 9.2*, each diagonally opposed pair of rods is electrically connected, and an oscillating voltage in the radio frequency range, combined with a direct current voltage is applied to the rods. One pair of rods has a positive voltage, or potential, while the other pair has a negative potential. Ions are introduced into the quadrupole mass filter at one end and travel along the axis of the quadrupole. As they pass through the quadrupole, they experience the dynamic field produced by the rods, and only ions of a certain mass-to-charge ratio have a stable trajectory through the filter. Ions with a different mass-to-charge ratio experience unstable trajectories and collide with the rods, getting neutralized and removed. By adjusting the voltages dynamically, the mass-to-charge ratio of the ions that can pass through the filter can be precisely controlled. This approach to mass filtering was a significant advancement in the field of spectrometry, allowing for more precise, efficient, and flexible ion separation and analysis.

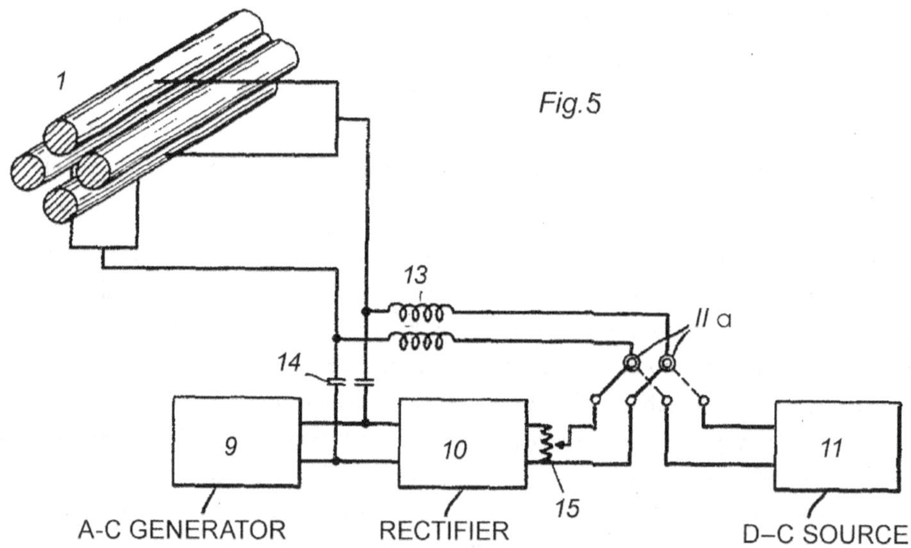

Figure 9.2: A diagram of the quadropole mass filter by Paul et al. - United States Patent and Trademark Office, Public Domain, https://commons.wikimedia.org/w/index.php?curid=3929798

However, Paul's innovation did not end with the quadrupole mass filter. Continuing to explore the potential of the dynamic field, Paul realized the combination of radio frequencies with the static charge could be manipulated to confine ions in three-dimensional space. Ions would enter one end of the quadrupole and, after filtering, could be suspended in space, stable for long periods of time in a kind of trap. At the time, there was a growing interest in precision measurements in atomic physics, particularly for fundamental constants and atomic weights, and the ability of Paul's trap to confine and study individual ions for extended periods was essential for such measurements. Paul would go on to be awarded a Nobel prize in 1989 for his contributions to the field.

At this point the story of ion trapping intersects with Isidor Rabi. In addition to the work discussed in the previous chapter, measuring nuclear magnetic resonance, Rabi also played an integral role in the scientific discoveries upon which all modalities of qubits depend. After some years working with his molecular beam magnetic resonance (MBMR) setup, Rabi suggested that his device might be used as a clock. The idea of using quantum mechanics to provide time keeping was not Rabi's originally, having been proposed by Scottish physicist James Clerk Maxwell in his *Treatise on Electricity and Magnetism,* writing in 1873, "a more universal unit of time might be found by taking the periodic time of vibration of the particular kind of light whose wave length is the unit of length."[4] Like all great theorists, Maxwell needed a brilliant experimental physicist to come along before his idea could be realized. After Rabi's suggestion in 1945, a project was initiated at the National Bureau of Standards (NBS), led by Harold Lyons, to use MBMR for timekeeping. In 1949, Lyons' efforts created a clock based on the absorption of microwaves by ammonia molecules, the first of its kind.[5] Called the NBS-1, the clock used the incredibly unwavering vibration of the molecules in the way Maxwell had suggested, and was accurate to within 1 second every 3000 years.

By 1955, advances in MBMR allowed the creation of clocks based on individual atoms at both the NBS and the National Physical Laboratory in the UK. These cesium clocks, named NBS-2 and Caesium I, respectively, achieved much greater accuracy than the ammonia-based NBS-1, gaining or losing 1 second only every 20,000 years. These devices represented quantum physics-based technologies with significant practical applications in everything from telecommunications to space exploration. Most importantly, it enabled a revolution in navigation — the Global Positioning System would not be possible without the accuracy of atomic clocks. In the physics community, atomic clocks brought about a new competitive drive — who could deliver another order of magnitude in precision?

One of the groups contributing to the drive towards greater precision was that of Norman Ramsey. Ramsey had been one of Rabi's collaborators on the MBMR experiments at Columbia before WWII, and had taken a position at Harvard after the war, where he continued to work with molecular beam experiments. In the early days of atomic timekeeping, the Ramsey group created what was known as the "separated oscillatory fields method," in 1949. This addition of another oscillating field separate from the original field, contributed substantial additional accuracy to MBMR measurements. By the mid-60's, a PhD student named Dave Wineland was completing his degree under Ramsey's mentorship.

Under Ramsey's direction, Wineland worked on precise measurements of the magnetic interactions between the nucleus and the electrons in orbit around it. The measurements were done using masers, devices that Ramsey had helped develop, a predecessor of lasers which emitted focused microwaves rather than visible light. This time with Ramsey would ignite in

Wineland a complete fascination with high resolution spectroscopy and precision that would carry through the rest of Wineland's career[6].

After earning his PhD in 1970, Wineland went to the University of Washington to join Hans Dehmelt's group, who were working on trapping ions in very similar ways to Wolfgang Paul, with whom Dehmelt would eventually share a Nobel prize in 1989. Wineland was initially drawn to Dehmelt's work by the advantages provided for the hyperfine measurements Wineland had been carrying out with the maser at Harvard. However, once in Washington his work instead focused on magnetic traps, known as Penning traps after Frans Michel Penning whose work at Phillips Research in the 1930s inspired their design.

After five years with Dehmelt, he joined the Time and Frequency Division at National Bureau of Standards, where he was tasked with helping make the NBS-6 clock operational, seen in *Figure 9.3*. This was a next generation cesium ion beam clock, with several advancements creating a new level of precision. The ions produced in the oven at one end travelled through the vacuum chamber, passing through a microwave cavity to the other end, where a detector counted the atoms in each of two energy states after interacting with the microwaves. Despite its increased accuracy, the ions in NBS-6 operated at room temperature, and therefore were limited to traveling close to the speed of sound, at an average of 300 meters per second, which made them susceptible to time dilation, per Einstein's special theory of relativity.[7]

Figure 9.3: David Wineland with the NBS-6 atomic clock. Credit: David Wineland/NIST https://www.nist.gov/nist-and-nobel/dave-wineland/context-discovery-dave-wineland

While in Washington, Wineland had written a paper with Dehmelt proposing a method to cool trapped ions using laser light. Atomic physicists had known for quite some time that higher levels of precision in their clocks would require cooling, but requiring a heat source that could produce a stream of charged atoms traveling at 300 meters per second did not lend itself to any normal method for cooling. With NBS-6 now online, Wineland began to work on the problem of laser cooling, returning to the trapped ion regime from his prior work.

Despite NBS' lack of focus on basic research, Wineland's group leader, Helmut Hellwig, believed the laser cooling might be valuable for the future of their timekeeping devices, so Wineland and some colleagues began work in the fall of 1977, using magnesium ions in a Penning trap. As is often the case in the sciences, another, unrelated effort was working in parallel towards a similar goal. The group, led by Peter Toschek at the University of Heidelberg, had been joined by Dehmelt, who was on sabbatical from the University of Washington. Their work used a Paul trap with a barium ion, however, both groups had the same goal: to use lasers to cool their ions to as low a temperature as possible.

The cooling effect relied on the fact that every atom has a very specific natural resonance frequency. A laser with a frequency that matches that of a particular atom will, when aimed at that atom, will cause it to either emit or absorb a photon, changing its energy state. The absorption of a photon from the laser also imparts a change in the motion of the atom, increasing momentum in the opposite direction of the laser. Emitting a photon occurred in a random direction, and so had no net effect on the atom's momentum. The subtle trick dreamt up by Wineland and Dehmelt was to tune the laser just below the atom's natural frequency. In the same way that the Doppler effect makes sounds coming towards you shift up in frequency, when the atom's moved towards the laser, the laser's frequency would shift up to the atom's resonant frequency, a photon would be absorbed, and the atom would be kicked in the opposite direction and emit a photon, diminishing the overall energy of the atom. Over the course of many such frequency shifts, the atom would be cooled.

In the fall of 1978, both groups had achieved the laser cooling effect on their trapped ions, with their respective papers reaching the scientific journal's offices within one day of each other.[8] Laser cooling immediately became a major area of research, with some atomic physicists exploring its applications on ions and others using it on neutral atoms, with both Paul and Penning traps being used experimentally, based on the desired goals. At NBS, the Time and Frequency Division produced its first laser-cooled atomic clock in 1984. Using a beryllium ion in a Penning trap, Wineland and his colleagues achieved comparable precision to the best cesium beam clock, and they were just getting started.[9] Thus, the story of Rabi's molecular beams and Paul's and Penning's traps finally intersected in Boulder in the 1980s, setting the stage for the experimental realization of qubits made of atoms.

One group using lasers on neutral atoms was led by Carl Wieman at University of Colorado Boulder. In Wieman's lab, gases made up of non-charged atoms were held in a vacuum chamber, where laser cooling was hoped toproduce temperatures low enough to create Bose-Einstein condensates. This state of matter occurs when atoms are super cooled and condense to their lowest energy state. At that point their quantum state becomes indistinguishable from one another, providing a macroscopic stage on which one may study quantum mechanical effects. Predicted by the theoretical work of Satyendra Nath Bose and Albert Einstein in the early 20th century, Bose-Einstein condensates required conditions of near absolute zero temperatures that had not yet been achieved, but in the 80s physicists believed they would soon have experimental results. One member of Wieman's group was a PhD student named Christopher Monroe, who, upon completing his degree, would join Dave Wineland's team.

The National Bureau of Standards had been renamed the National Institute of Standards and Technology (NIST) in 1988, and Wineland had founded the Ion Storage Group to continue the work begun by the laser cooled ion trapping experiments of the late 70s. Once at NIST, Monroe began to help with efforts to entangle ions within the same trap. Each entangled ion would multiply the accuracy of the device when used as a clock by augmenting the frequency of the system. Ten entangled ions would create tens times the "run rate" of the clock, adding yet another order of magnitude to the resulting precision of its timekeeping. By 1990 they had also demonstrated putting trapped ions into superposition of both its ground, or low energy, and excited, or high energy states. This was accomplished using very carefully tuned and timed laser pulses after cooling the ion as much as possible.

What had begun with streams of atoms aimed at detectors in Rabi's lab was now fully supplanted by the trapping model, where small collections of atoms could be held suspended in three-dimensional space. This incredible feat required careful preparation of the fields used to create the capture zone, and not a small amount of luck to find the ion once it was in there. Monroe and the other staff members could sometimes hunt for weeks before finding the tiny flash of emitted light that signified the presence of the ion. Once located, however, the system was remarkably stable, with the ion's state capable of being preserved for hours or even days at a time. It made for an incredibly powerful new platform for physics experiments.

In 1994, the International Conference on Atomic Physics was held at the University of Colorado Boulder,[10] and a guest lecture was delivered by Artur Ekert, a physicist from the UK who had done his PhD at Oxford under David Deutsch. Ekert's thesis had described a method for using Bell's inequality as a mechanism for quantum encryption, which he arrived at independently. In the pre-Internet days research papers were distributed very inefficiently, so Ekert only learned of prior quantum encryption work by Bennett and others once the similarities occurred to Deutsch, and he mentioned them to him. Ekert became a kind of quantum evangelist, giving talks, hoping to inspire and recruit new enthusiasts to the field, and especially hoping to rope experimentalists in, so that they might build something that worked for theorists to use.

At NIST, Ekert gave a talk on the theory of quantum computing, explaining its potential power and also that no one knew how to build a quantum computer, or even if such a thing were possible. In the audience were two theorists from Europe, Peter Zoller and Ignacio Cirac. Zoller had taken a professorship at Boulder earlier that year, and Cirac, who had done part of his PhD at the University of Innsbruck with Zoller, was a visiting professor from University of Castilla-La Mancha in Spain. Inspired by the challenge, Zoller and Cirac immediately set about working on a scheme to create a physical qubit. Working in quantum optics at Boulder and at the Joint Institute for Laboratory Astrophysics (JILA), a facility where academic researchers collaborated with NIST scientists, it was natural for them to think of ion traps and lasers. In fact, through JILA they were familiar with Wineland and the work his group was doing.

In a few months, they had a draft paper, *Quantum Computation with Cold Trapped Ions*[11], which they submitted to the journal Physical Review Letters, and shared with Wineland. The journal referees initially rejected their work, saying it was impossible to build such a device. Wineland and Monroe, however, were galvanized into action. Their work on improving the control of the ions for time keeping devices was so well aligned with Cirac and Zoller's proposal it seemed almost trivial for them to implement. In what may be the most efficient collaboration between theorists and experimentalists, Wineland and Monroe demonstrated a gate operation on an ion in a Paul trap just a few months after the paper was published, in 1995. Trapped ion qubits had arrived.[12]

As much as it was a huge milestone, in practical terms, the initial result out of NIST was modest. While Zoller and Cirac had laid out proofs that trapped ions could support a universal set of gates capable of supporting quantum algorithms, the Wineland groups' experiment had simply applied a two-qubit gate, a controlled-NOT, to a single ion. The internal hyperfine states of the atom represented one qubit, and external motional states represented the other. However, it was a solid basis for continuing work, and by 1997 Wineland's group had demonstrated a CNOT applied to two ions in single file inside a trap. In multiple ion devices, the hyperfine states, or spin states, of the ions are used as the two level systems of an individual qubit. Two qubit gates are applied by interacting with each atom's external motion, called motional state, with the two ions' motional states interacting as a single "phonon," or quantized vibrational state.

Other groups, inspired by both the theory paper from Cirac and Zoller and the results out of NIST, turned to building experimental qubits out of ion traps. Back at the University of Innsbruck, Cirac and Zoller found experimental partners in Rainer Blatt's group, while Hartmut Haffner led a group at the Max Planck Institute for Quantum Optics in Munich, Germany. Back in the US, a group at Caltech under the leadership of H. Jeff Kimble was also conducting experiments with trapped ions, though their work focused mostly on the extension of Feynman's quantum electro dynamics (QED) to the context of the microwave cavity in the trap. This work, known as cavity QED, was to play an important role in the development of other types of qubits, especially those made from superconducting circuits.

The theorists weren't done with their contributions, either. While the controlled-NOT gate operation proposed by Zoller and Cirac had been made to work at Innsbruck in 2003,[13] it was very challenging to do reliably. It required the two ions to be held in their joint motional ground state during the operation, in other words sharing the lowest possible energy of their collective motion, and also required a complex sequence of one- and two-qubit actions performed by the device's lasers. This required extreme precision with the device and made the entire enterprise quite sensitive to noise, thermal or otherwise. Any disturbance or imprecision would cause the qubit to undergo "decoherence," a transition from quantum to classical behavior, and any computation would be lost. The problems would be exacerbated in larger systems, where maintaining a large number of ions in their joint ground state would increase exponentially in difficulty.

Two physicists at Aarhus University in Denmark, Klaus Mølmer and Anders Sorensen, suggested another approach that was insensitive to the vibrational state, and published a paper in 1999 titled *Multiparticle Entanglement of Hot Trapped Ions*[14] describing their technique for operating on all ions in a one-dimensional trap at once. An Australian researcher, Gerard Milburn, suggested some enhancements[15] and by 2000, a solid theoretical model for what came to be called Mølmer-Sorensen gates had been formed. Wineland's group performed an experimental validation that same year at NIST, operating on four trapped ions with reasonable fidelity and relative ease compared to Cirac and Zoller's method. Mølmer-Sorensen gates completely supplanted Cirac-Zoller gates, and dominate the operations performed on ion trap quantum computers to the present day.

Very soon after, Ike Chuang, having abandoned liquid state NMR qubits when he left IBM, was setting up his new lab at MIT with trapped ion qubits in 2001. Chuang had earned his PhD at Stanford with Yoshihisa Yamamoto as his advisor, and was a co-author on the paper laying out the proposal for an atom and photon based quantum computer back in 1988.[16] In that sense the pivot to ion traps was almost inevitable, as both atoms and photons were utilized in this scheme as well, and had the advantage of having a physical manifestation. Chuang became involved in improving trap design itself, contributing to efforts to realize a planar ion trap, with the quadrupole and its electrodes laid out on a two-dimensional "chip" of sorts.

This technique, often called "surface electrode traps," was an important innovation, allowing the use of CMOS, or "complementary metal–oxide–semiconductor" technologies for fabrication, the same as classical computer chips. Forward-looking researchers, anticipating the engineering challenges of creating systems with enough qubits to do useful work, wanted to create a fabricated platform that not only would accommodate the trap but, eventually, the optics as well. They imagined this architecture paving the way to systems with hundreds, thousands and even millions of trapped ion qubits.

After the disappointment of NMR's limitations, it would have made for an excellent *denouement* for trapped ions to have proven to be unambiguously the best technology for the job. However, that did not seem to be the case. Despite the surface electrode trap, advances in scale came slowly, and the size of the systems that could be built in the 2000s stalled out around five qubits. Enthusiasm began to wane for trapped ion qubits, and a notable paper laid out the case that all two qubit gates had been tried and there wasn't much more to learn.

This conservatism may have been a by-product of the world of atomic time keeping, with its emphasis on precision. After all, the devices being built at NIST were providing the accurate, reliable timing on which critical systems around the world relied. While ion trappers were proud of the long coherence times and high gate fidelities of the devices they were building, no-one had any compelling strategy to get to systems with the scale and execution times required to outperform classical computers. The gate times were slow, for one thing, taking microseconds for each operation. At those speeds, a trapped ion computer's operations would be orders of magnitude slower than the gates on classical systems. The theoretical advantages of quantum algorithms had mostly to do with making intractable classical problems solvable, so if the system took an impractical amount of time to solve a problem it offered no practical solution at all.

More daunting was a method to achieve the scale required for practical usefulness. The ions in a Paul trap line up single-file in a single dimension. Penning traps could be two-dimensional but had significant hurdles to overcome before they could be used for qubits and no-one was sure if that would even prove to be possible. This one-dimensional trap had a hard upper limit of about thirty ions, with larger numbers of ions creating instability and degrading performance. The only way to get into the range of thousands of qubits seemed to be to couple traps one to another using photons. This was feasible but again, faced significant headwinds and major challenges. Some were going to require advances in engineering, others required novel science. Others simply were not yet identified.

By 2010, the funding agencies who had been enticed and terrified by the potential for Shor's algorithm to crack all codes were taking stock fifteen years in. The qubit counts were stalling out under 10, several orders of magnitude smaller than would be needed for Shor's. Some influential voices in the researcher community were casting doubt on the whole enterprise. Even if it proved possible, the road ahead was going to be long. The clock community continued to improve trap-based devices to gain ever more orders of magnitude of precision. Experimental physicists continued to use the technique for the study of atomic physics and quantum mechanics. But the vision of using nature's perfect qubits in large scale engineered computing devices seemed elusive. Perhaps another approach was needed, one that expanded the scale of the quantum world to something closer to our own. Instead of atoms emitted from an oven and caught in mid-air, these qubits were constructed out of superconducting circuits held at colder temperatures than occur anywhere in nature.

References

1. "Quantum Computing with Molecules." Scientific American, 1998
2. https://en.wikipedia.org/wiki/J._J._Thomson
3. https://www.rigb.org/explore-science/explore/blog/subatomic-science-jj-thomsons-discovery-electron
4. Achard, F. (2005), "James Clerk Maxwell, A treatise on electricity and magnetism, first edition (1873)", Landmark Writings in Western Mathematics 1640–1940, Elsevier, pp. 564–587, doi:10.1016/b978-044450871-3/50125-x, ISBN 9780444508713
5. https://sova.si.edu/record/NMAH.AC.0701
6. https://www.nobelprize.org/prizes/physics/2012/wineland/biographical/
7. https://www.nist.gov/nist-and-nobel/dave-wineland/context-discovery-dave-wineland
8. Wineland, D. J. (2013). Nobel Lecture: Superposition, entanglement, and raising Schrödinger's cat. Reviews of Modern Physics, 85, 1103–1114. https://doi.org/10.1103/RevModPhys.85.1103
9. https://journals.aps.org/prl/abstract/10.1103/PhysRevLett.54.1000
10. International Conference on Atomic Physics et al. Atomic Physics 14 : Fourteenth International Conference on Atomic Physics Boulder Co 1994. American Institute of Physics 1995.
11. https://journals.aps.org/prl/abstract/10.1103/PhysRevLett.74.4091
12. Demonstration of a Fundamental Quantum Logic Gate C. Monroe, D. M. Meekhof, B. E. King, W. M. Itano, and D. J. Wineland Phys. Rev. Lett. 75, 4714 – Published 18 December 1995
13. Schmidt-Kaler, Ferdinand; Häffner, Hartmut; Riebe, Mark; Gulde, Stephan; Lancaster, Gavin P. T.; Deuschle, Thomas; Becher, Christoph; Roos, Christian F.; Eschner, Jürgen; Blatt, Rainer (2003-03-27). "Realization of the Cirac–Zoller controlled-NOT quantum gate". Nature. 422 (6930): 408–411.
14. Mølmer, Klaus; Sørensen, Anders (March 1, 1999). "Multiparticle Entanglement of Hot Trapped Ions". Physical Review Letters. 82 (9): 1835–1838.

15. Milburn, G. J. (August 12, 1999). "Simulating nonlinear spin models in an ion trap". arXiv:quant-ph/9908037
16. K; Yamamoto, Y (1988). "Quantum mechanical computers with single atom and photon fields". International Quantum Electronics Conference.; Chuang, I; Yamamoto, Y (1995). "Simple quantum computer". Phys. Rev. A. 52: 3489–3496.

10

Man-made Atoms

Do the flux-qubit experiments... "probe the quantum-classical interface" as is not infrequently claimed in the abstracts or conclusions of the relevant papers? I am afraid that the situation here is asymmetrical: If in these experiments one were to find evidence for a breakdown of the QM predictions ...this would be a major contribution ... However, if we keep finding (as we have so far) the behavior predicted by the straightforward extrapolation of the QM formalism to this level, I am afraid we will really have learned nothing.

— *Tony Leggett*[1]

Macroscopic sub-atomic machinations

The story of superconducting qubits begins with the phenomenon of superconductivity itself. When electrons travel through a material, they encounter resistance, a sort of friction against their forward movement that causes some energy to dissipate in the form of heat and other byproducts. Resistance is why an electrical device will feel warm to the touch when plugged into a power supply. In the late 1950s, a trio of physicists at the University of Illinois Urbana-Champaign collaborated on research into the conditions under which resistance disappears and a material behaves as a superconductor. John Bardeen, Leon Cooper, and their grad student Robert Schrieffer, were working with materials that, when cooled to very low temperatures using liquid helium, would cause electrons to be attracted to one another and form pairs. These are now called Cooper pairs, and once they have formed all the pairs of electrons move in correlation with one another, with absolutely no resistance. A current flowing through a superconducting circuit will go on forever with no energy dissipation, even after the source of the current is disconnected, so long as the conditions for superconductivity remain. The formalization in 1957 of their understanding of superconducting described by the trio took on their last names, Bardeen-Cooper-Schrieffer, or the BCS theory, for short, for which they earned a Nobel Prize in 1972.

BCS naturally set off a superconducting craze in the physics community. There were fantastic ideas about levitating trains, lossless energy transmission, the hunt for materials that might be superconductors without super cooling, and the less fanciful desire to understand the further implications of these new insights into quantum mechanics. In 1962, a 22-year-old PhD student at the Mond Laboratory of Physics at Cambridge named Brian Josephson predicted that under certain conditions Cooper pairs of electrons would form across an insulating gap in a circuit fabricated from superconducting material. This prediction was widely regarded as untrue and foolish. The young graduate student found himself debating John Bardeen himself, in a public forum. Experimental verification came, however, the very next year, thanks to two other British physicists, Philip W. Anderson and John M. Rowell. The device they built consisted of a superconducting circuit punctuated by an insulator, as seen in *Figure 10.1*, which should have prevented electrons from traveling through the circuit, but didn't. The electrons in this device, now known as a Josephson junction, formed Cooper pairs on either side of the insulator, effectively tunneling through the nonconductive material. The theory and proof of electron tunneling earned Josephson vindication in the form of a shared Nobel Prize in 1973.

Though the Josephson junction was initially just an experiment designed to validate the hypothesis of electron tunneling, it inspired enthusiasm as a device unto itself. It was easy to fabricate — add an insulating barrier of something like aluminum oxide a few nanometers thick to a simple circuit made of material that was a superconductor at very low temperatures. Most importantly, though, it exhibited quantum coherence at a macroscopic level. The electrons in Cooper pairs are held together by a force which counteracts the normal repelling charge one would expect of two like-charged particles. This attractive force extends further out from one electron than the charge force, so far, in fact, that it can bridge the insulating gap in the Josephson junction. Across the insulating gap a Cooper pair maintains its connection, and current flows through the broken circuit. The wave function that describes that Cooper pair applies to *all* of the Cooper pairs in the superconducting circuit, and so the phase coherence, the result of the force exerted from one electron to the other across the insulating gap, is present in the entire circuit.

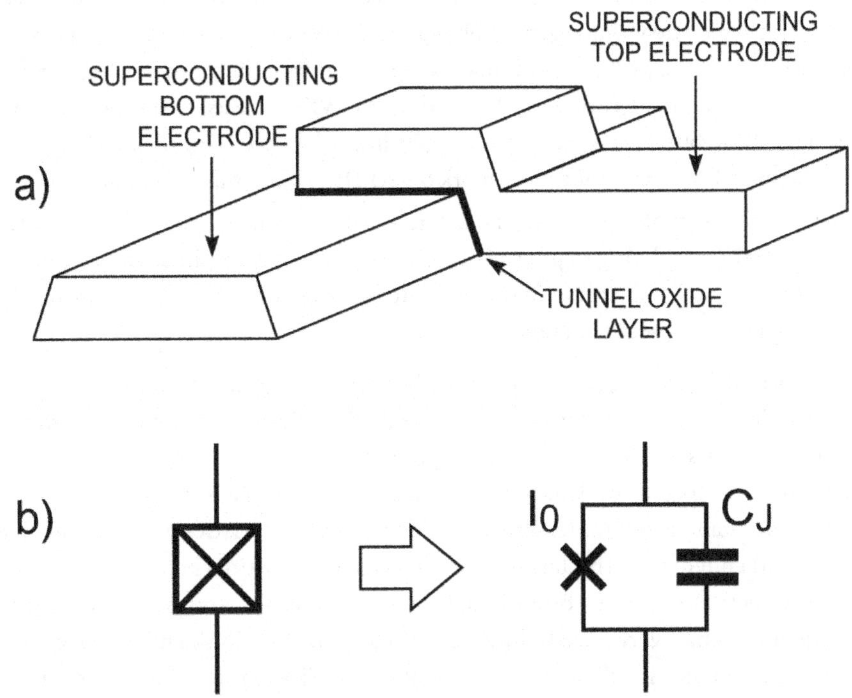

Figure 10.1: a) is a typical design for a Josephson junction made with two superconducting films separated by a non-conductive oxide layer; b) is the schematic representation of a Josephson junction https://arxiv.org/abs/cond-mat/0411174

When studying the impossibly small scale of quantum mechanics, any opportunity to increase to merely incredibly small is very welcome, and so a wave of experimentation with Josephson junctions followed. In addition to adding a powerful new tool to the experimentalist's lab, there was also interest in possible practical applications of the scientific breakthrough. At Ford Research Labs, researchers exploited the electromagnetic characteristics of the Josephson junction to build what became called a SQUID, for Superconducting QUantum Interference Device. The electron tunneling is extremely sensitive to the magnetic flux created by the current in the superconducting circuit, and so, properly instrumented, a junction can become a very sensitive magnetometer, able to detect minute changes in magnetic fields.

Two academic researchers were particularly interested in SQUIDs, John Clarke, who earned his PhD at Cambridge before joining the physics department at UC Berkeley in 1969, and Tony Leggett, who studied at Oxford before spending the late 60s at UIUC with Bardeen and others exploring superconductivity and its effects. Clarke was drawn to the Josephson junction as a tool, seeming to have an intuition that it was more than just an experimental demonstration of theory and might be put to other uses.

Leggett, in addition to the substantial contributions to the theory of superconductivity and superfluidity, for which he would earn a Nobel prize in 2003, began encouraging the physics community to explore the possibility of macroscopic quantum mechanical phenomena in SQUIDs and other Josephson tunneling-based devices. While at Sussex, Leggett's initial work had focused on NMR, but by the early 80s, Leggett had taken the John D. and Catherine T. MacArthur Chair at UIUC and continued work on SQUIDs, studying "the quantum superposition of states that are by some reasonable criterion macroscopically distinct." [2] Leggett's ideas were tantalizing to physicists because what they implied was that it might be possible to engineer artificial atoms that would provide the opportunity to study behaviors normally hidden from view at the subatomic scale.

While neither Clarke's nor Leggett's work directly referenced quantum computing at this stage, both had indirect, outsized influence on the development of the field through their work on Josephson junctions and SQUIDs. An undergraduate at the University of California Berkeley named John Martinis took a condensed matter course from Clarke in his fourth year and was captivated by the challenge of understanding the noise characteristics of Josephson devices, stuck around and ended up with Clarke as his PhD advisor. Martinis had a background tinkering with electronics that he brought to his study of physics, and was attracted by the fact that the frequencies that were used to interact with SQUIDs was in the microwave range, a well-understood domain. This observation would play a critical role in the future of the technology for computation.

At conferences, Martinis encountered Leggett's study of macroscopic quantum phenomena and his interest in Josephson junctions deepened. During his PhD he was joined at UC Berkeley by another SQUID enthusiast, Michel Devoret, doing a postdoc with Clarke after earning his PhD from the Commissariat à l'Energie Atomique (CEA) in Saclay, France. At CEA, his advisor, French physicist Anatole Abragam had deep experience with nuclear magnetic resonance devices, which equipped Devoret exceptionally well for the experimental work in Clarke's group, especially due to his experience with dilution refrigerators. Martinis and Devoret seemed to complement each other well, and accelerated the progress in SQUID experimentation enormously. Martinis hit upon the idea of adding a second, resonant Josephson junction for readout, achieving some degree of isolation from noise in the device. Most critically Devoret and Martinis supplanted the qualitative analysis typical of the work done to that point with more rigorous, quantitative data collection and analysis.

In 1985, Devoret, Martinis and Clarke published *Measurements of Macroscopic Quantum Tunneling out of the Zero-Voltage State of a Current-Biased Josephson Junction,* which presented analysis of the microwave signals collected from a Josephson junction that validated the quantum mechanical behavior present in the device. Lacking the theoretical foundation for quantum computing, there was no way at this stage to recognize the suitability of the system to support computation. Martinis recalls a conference held around the time of the Clarke, Martinis and Devoret paper, where, in addition to the topic of macroscopic quantum tunneling in the superconducting experiments at UC Santa Barbara, the last speaker was Richard Feynman. Feynman spoke on the topic of quantum computing, which strongly implied that some conference organizers' saw the possibility of an intersection of the two topics in the future.

Leggett's work at UIUC continued to explore macroscopic quantum tunneling and coherence, and to popularize the notion that Josephson junction-based systems could provide a window into the dynamics usually relegated to the sub-atomic scale. One researcher with NEC's Fundamental Research Group in Japan, Yasunobu Nakamura, who had been working with single electron transistors, read Leggett's work and was intrigued. Shifting his work to Josephson junctions, he and his supervisor, Jaw-Shen Tsai began building superconducting devices. Within a year they gained access to a dilution refrigerator and began cooling their aluminum devices down to the millikelvin range.

Nakamura, Tsai and their collaborator Yuri Pashkin rapidly assimilated learnings from various groups who'd worked with Josephson junctions like Clarke, Martinis, and Devoret, as well as more recent enthusiasts such as Hans Mooij at Delft in the Netherlands and Daniel Esteve at CEA. Nakamura also drew on work he'd encountered in single electron transistors from a variety of Russian researchers, one of whom, Nasalov, had written a paper on quantum dots, an application of single electron transistors, which contained formulae that described the dynamics in the device being built at NEC. Traveling to Europe for the first time in 1996, Nakamura met many of these colleagues, hitting it off with Devoret in particular who passed the word through the community that Nakamura was doing interesting work, leading to meetings with Delsing at Chalmers, Esteve at CEA, Mooij at Delft and others. When Nakamura met with Nasalov, however, the Russian had trouble understanding where Nakamura had found his formulae, even when Nakamura explained they had come from his own published work.

After this fruitful trip Nakamura returned to Japan and shortly thereafter demonstrated superposition of the two charge states in their device. One Josephson junction on the device was designed to facilitate Cooper pair tunneling, creating what is referred to as a "quasi-particle," or particle-like behavior out of the interaction between the two electrons in the pair. On the other side of the device, another junction with a high voltage bias was designed for electron tunneling. The junctions were linked such that, when a Cooper pair tunnels into the reservoir between them, two electrons can escape across the other junction. Without the Cooper pair present, the electrons are blocked. Monitoring the current across the electron tunneling junction therefore serves as a very sensitive probe for the presence of the quasi-particle. Careful measurements showed that the two charge states, the Cooper pairs present and not present, could be put in superposition. This, finally, was the two-level system that would be required to create a qubit using a Josephson junction.

Published in 1997 in Physical Review Letters, Nakamura shared those results with his colleagues in Europe, and sent an email to Tony Leggett, asking if he agreed if it was evidence for macroscopic quantum coherence that he had proposed. Leggett responded that he didn't consider a single Cooper pair "macroscopic," as he had been talking about large numbers of electrons acting in a quantum mechanical way at a larger scale. The debate over what "macroscopic quantum coherence" meant exactly, remained a side show to the practical progress on superconducting qubits for many years.

Strictly speaking the proposal that Nakamura, Pashkin and Tsai had published would not produce a qubit at this stage. In fact, the 1997 paper didn't mention quantum computing at all. The motivating force was simply curiosity - could they demonstrate superposition in their system? Others who they had been sharing results with were thinking about quantum computing, however, and a variety of papers in 1997 and 1998 from researchers at CEA, Karlsruhe and Delft all pointed the way to developing proper qubits from these superconducting devices.

In 1998, using an expensive high frequency pulse generator borrowed from an optical communications group at NEC which was capable of pulses of just 100 picoseconds, or trillionths of a second, and despite skepticism from their peers, Nakamura, Pashkin and Tsai were able to demonstrate coherent control of a two-level system. Their device had a coherence time of only one or two nanoseconds, but it was capable of quantum coherence at a macroscopic scale — a qubit made out of an artificial atom. Twelve years after the Berkeley result, the superconducting qubit was born.

The paper marks the beginning of a period of extremely rapid progress and wildly creative innovation among experimentalists working towards superconducting qubits. A design component originating in Esteve's lab at CEA, an iteration of which formed the basis of the Nakamura, Pashkin and Tsai device, was refined and earned the name "Cooper pair box" around the same time as the Nakamura paper was published. This box optimized the superconducting "island" separated from a "reservoir" by a Josephson junction so that precise control of the charge state of the whole system became more easily achievable, and formed the basis of much of the subsequent innovation.

By 2001, Nakamura, Pashkin and Tsai had refined their device and published another paper measuring a Rabi oscillation - a periodic transition between two energy states in resonance with a driving electromagnetic field. Around the same time, Mooij and others at Delft created what they called a flux qubit, where the two levels were represented by the electromagnetic state of the system rather than the charge, which presented a different set of strengths and weaknesses. Esteve's group at CEA came up with something they called a "quantronium" qubit with an enlarged superconducting island in an attempt to make the device less susceptible to charge noise. The neologism formed by tacking "-onium" on the end of a physics word followed a tradition in the field that began in the middle of the 20th century when work on anti-matter began referring to "positronium," to refer to the bound state of an electron and its antiparticle, a positron. With quantronium, Esteve resurrected the convention and started a trend in the superconducting community to name designs in a manner that sometimes sound better suited for comic books.

In 2002, Devoret left CEA and took a role at Yale University, where he began collaborating with Rob Schoelkopf and Steve Girvin. Yale rapidly became a major center of gravity for superconducting qubits, beginning with a major theoretical breakthrough just two years after Devoret arrived. Quantum electrodynamics was Richard Feynman's contribution to the Standard Model of particle physics, providing a complete model to describe the interactions between photons and charged particles like electrons. Often referred to as "QED," it was the work for which Feynman was awarded a Nobel prize in 1965. Working with trapped ions, Serge Haroche had updated the theory for that specific context, producing work in 1996 known as "cavity QED." Now, the Yale Quantum Institute team, including Alexandre Blais, Andreas Wallraff, and Ren-Shou Huang working under the supervision of Schoelkopf and Girvin worked on a further update to QED, which they called circuit QED.

The paper, published in 2004, laid out a blueprint for understanding and improving on the superconducting qubit.[3] The framework presented a one-dimensional "cavity" in the form of a superconducting circuit to stand in the place of an atom in cavity QED. The dynamics of this artificial atom were thoroughly explored as a means of realizing qubits, including addressing issues of noise and coherence and advocating a focus on the strong coupling between the Cooper pairs in the circuit and the microwave photons of the control signals as a means to address noise. The authors also described in detail the design of these devices, and the means for applying gates and reading out qubit state.

Following on from that theoretical baseline, the researchers at Yale and their collaborators went on to produce the two dominant species of superconducting qubits, creating a reference design and theory for charge qubits in 2007 and for flux qubits in 2009. The comic book naming trend continued, as the flux qubit earned the name "fluxonium," and the charge qubit was dubbed the "transmon." Transmon was supposedly a portmanteau for "transmission line shunted plasma oscillation qubits," but that seems like the first example of quantum computing retcon. Whatever its true origin, the naming of the transmon contributed a new suffix to naming trends, and has spawned many further variations such as unimon, coaxmon, Xmon, and others.

The transmon qubit is designed to reduce sensitivity to charge noise, a common problem in earlier qubit designs like the Cooper pair box. The key innovation in the transmon is its use of a large shunt capacitor, which increases the qubit's coherence time while maintaining the non-linear characteristics essential for qubit operation. This stabilizing feature improved the overall performance of the transmon, making them competitive with flux qubits. Flux qubits, due to their use of electromagnetic flux as the means of storing and manipulating state, have reduced sensitivity to charge noise, giving them a natural advantage over charge qubits like the transmon, in addition to their faster operations thanks to higher energy gaps. However, the transmon's simpler design provided an easier path to overall improvements in system performance.

Experimentalists continued developing innovations such as three-dimensional resonator cavities that led to improved coherence times, all electronic control of qubits, and innovation around error correction, but the major contribution would remain the transmon and the fluxonium qubits. Transmons in particular simplified the design and fabrication of the devices sufficiently to make building qubits much more accessible, effectively leveling the research landscape and become the dominant design in labs around the world.

In parallel to the work on gate-based superconducting qubits, another way to use quantum tunneling had been developed in 1999 by two Japanese researchers, Kadowaki and Nishimori, called quantum annealing.[4] Annealing is the process in metallurgy where controlled cooling is used to achieve desired effects, for example making sheet metal less prone to cracking while being shaped. In 1983, computer scientists used simulated annealing to solve the "traveling salesman problem,"[5] and by the late nineties it was a standard heuristic tool used for similar applications. The idea of the quantum researchers was to use the Josephson tunneling to arrive at a solution more efficiently than classical methods.The advantage of quantum annealing is that gates are not needed, which lowers the level of engineering sophistication required of the overall system architecture. However, because annealing is a heuristic and not algorithmic, meaning at its core it is based on guesses that are not verifiable, there is no guarantee of arriving at the best final answer. However, the idea was attractive enough to launch the world's first quantum computing startup, D-Wave Systems, founded in Vancouver, Canada, shortly after the paper was published.

As work on gate-based superconducting devices progressed, so too did the work in D-Wave's R&D labs, and by 2011 the company made the first quantum system commercially available, the D-Wave One, with 128 superconducting annealing qubits. John Martinis, having spent time at NIST and now at University of California Santa Barbara, met the D-Wave team at a conference in Vancouver just before the commercial launch, in 2010. Martinis was drawn in by the engineering feat of shipping a quantum technological product, and, along with other researchers including Matthias Troyer, began working on characterizing the performance of the unique machine.

In 2013, D-Wave Two was launched, equipped with 512 annealing qubits, and a system was purchased by the NASA AMES lab, who had entered into a collaboration with Google to explore the potential applications of quantum annealing. The same year, Martinis, Troyer, and others published two papers[6] showing that the initial apparent performance gains achieved on the D-Wave system could be matched by clever classical simulation code primarily written by Troyer. This is the first example of what has become known as "dequantization," an awkward term for challenges to claims of quantum advantage with clever approaches using classical computing, such as the effort by IBM in response to Google's supremacy experiment.

While doubts were beginning to be cast on the performance of the annealing qubits, progress on gate-based qubits was continuing. Martinis' group at UCSB had a system with five gate-based qubits and were planning an upgrade to nine qubits, with hopes they could implement an error correction scheme called surface codes on the chip. This technical roadmap could feasibly deliver on the potential promised by the theorists like Shor and Deutsch a decade earlier, and between the papers and the working system, Martinis attracted the attention of the head of Google's quantum team, Hartmut Neven. After delivering a talk to the Google researchers, Neven asked Martinis if he and his group would consider bringing their project to Google, and Martinis agreed.

Meanwhile, the "Yale gang" in the northeast of the U.S. was having its own impact on the superconducting qubit landscape. Chad Rigetti, Jay Gambetta and Jerry Chow all emerged from the transmon efforts at Yale to find their own ways to IBM Research, where foundational work had steadily continued since the 80s. The company was now ready to make a serious investment in moving towards commercializing the nascent technology, and the transmon seemed the most suitable vehicle. Gambetta focused IBM's efforts, enlisting the expertise of the chip fabricators and the equipment at T.J. Watson and convincing the company to cease investment in other qubit modalities. Chow led the chip design team and Gambetta assumed control of the company's efforts to incubate a new business based on quantum computing.

Another tech giant had decided to enter the fray along with Google and IBM — Microsoft Research, led by mathematician Michael Freedman, had begun investing in a superconducting qubit effort in 2012, hiring Matthias Troyer as their lead theorist. Instead of duplicating the approach used at UCSB or Yale, they had landed on a novel approach to the modality, announcing that they would attempt to build something called topological qubits. Topological quantum states were pointed out to have significant advantages for fault tolerant computation by Kitaev, a famed researcher at Caltech. When a quantum system is in a topological state its information is preserved through noise that would destroy the state of a non-topological system. Making a qubit that could induce and maintain a topological state as part of the operation of its two-level system seemed like a shortcut to a fault-tolerant quantum computer, and in that context Microsoft's technical direction made sense.

However, to say that topological qubits were elusive would be a massive understatement. Theoretically, a variation on the Josephson junction architecture should be able to conjure up a very exotic quasiparticle called the Majorana Fermion, but no-one had to that point been able to achieve that result in a lab. Undaunted, Microsoft went ahead, hiring Leo Kouvenhoven, a Dutch researcher at Delft who had published a promising glimmer of something resembling a Majorana in 2012, and Charlie Marcus, an American condensed matter physicist working at the Niels Bohr Institute in Copenhagen and widely seen as the best fabricator of nanoscale materials. The hunt for a Majorana was on, and in Marcus and Kouvenhoven's labs as the researchers fabbed one test device after another, the quasiparticle proved as elusive as the physicist who lent it his name, Ettore Majorana, who had disappeared without a trace in 1938.

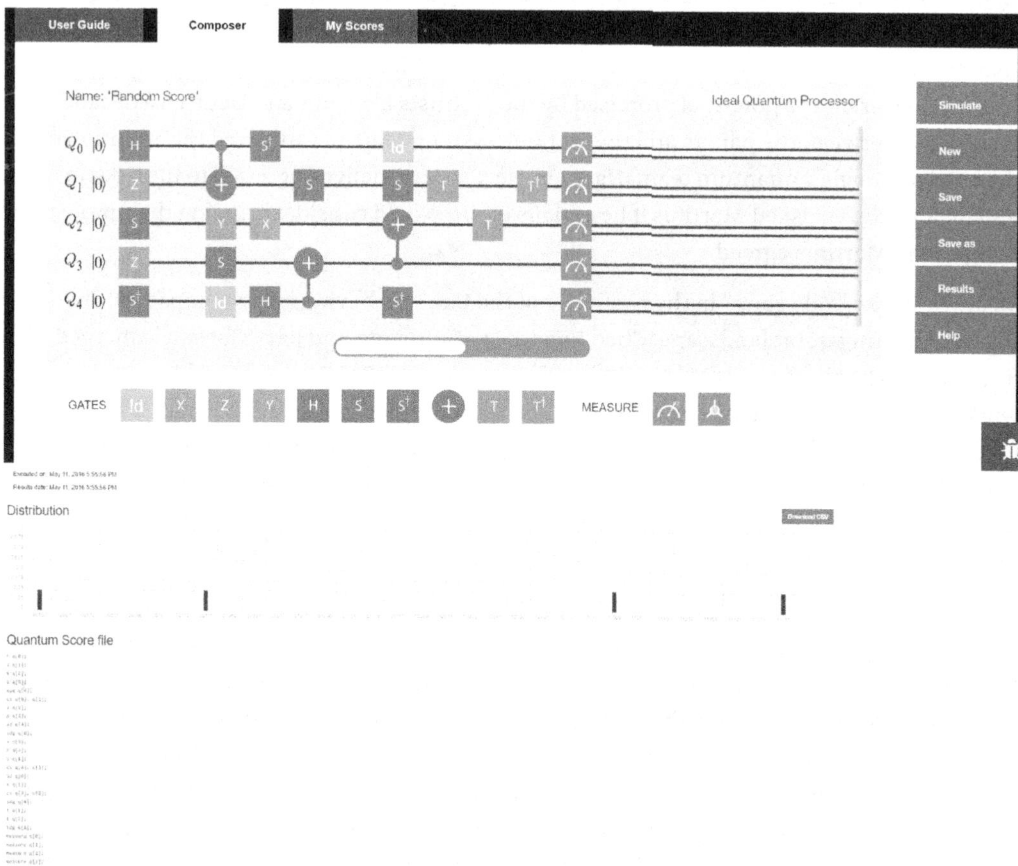

Figure 10.2: IBM's Composer interface for constructing quantum circuits

IBM's choice to focus on transmons quickly began to pay off, catching them up to superconducting efforts with their own five qubit system, and instilling the IBM Research team with confidence about their technological roadmap. The simplicity of their design made transmons easier to fabricate, and helped IBM Research bring to bear the decades of non-superconducting chip fabrication know how. They believed they could iterate on their design faster than their competitors, increasing quality and scale at a pace that could sustain progress towards commercialization. In May of 2016, IBM Research launched the IBM Quantum Experience[7], a website that provided access to a five-qubit machine.

Here for the very first time anyone with an Internet connection could use a visual interface, the Composer (seen in *Figure 10.2*), to drag and drop quantum gate operations on representations of the physical qubits in the device. The sequence of gates, strung along "wires" that represented the evolution of each qubit through time and resembled a musical staff, was referred to as a circuit. A quantum circuit could be constructed to represent a simple algorithm, and then be submitted via API call to run on the real device. It was intoxicating to imagine the symbolic representation of a simple algorithm being translated into a series of microwave pulses, then transmitted down wires leading into the dilution refrigerator at T.J. Watson, down past four Kelvin into the millikelvin zone, the "coldest place in the universe," as experimentalists like to point out, to ultimately poke and prod at the artificial atom coursing through the superconducting circuit, returning the signal that measured its quantum state.

It was also unlike anything any software developer working in 2016 had ever seen, more like the hand-crafted programs run on Von Neumann's computer in the 40s than even the most low-level programming of the modern era. If there ever was going to be a true quantum computing industry, it would need an entirely new kind of software developer possessing the requisite skills, too. Encouraged by the positive reception of the Quantum Experience, the next year IBM launched a commercial offering called the IBM Q Network. For an annual fee, the largest IBM customers could gain access to larger experimental devices not available to the general public, and more importantly, access to the IBM research scientists who could hold their hands through their first exploration into the unfamiliar territory.

Around the same time as the Q Network launch, IBM released an open source software developer kit (SDK), called Qiskit. The decision to release the SDK as open source was made by Gambetta after he and his lead software architect hastily coded up the first version over a weekend, anticipating the decision would get bogged down in IBM's sometimes bureaucratic processes. With the SDK, developers everywhere in the world could write code in Python, a very familiar high-level language with tens of millions of users, running quantum circuits via the Quantum Experience APIs and getting the results right on their laptops. IBM Research had radically democratized access to these exotic devices, making the transmon the de facto standard of what now seemed to be the next big information technology revolution.

References

1. Annu. Rev. Condens. Matter Phys. 2020.11:1-16.
2. nu. Rev. Condens. Matter Phys. 2020.11:1-16. Downloaded from www.annualreviews.org Access provided by 2601:645:8400:2
3. Alexandre Blais; et al. (2004). "Cavity quantum electrodynamics for superconducting electrical circuits: An architecture for quantum computing". Phys. Rev. A. 69 (6): 062320. arXiv:cond-mat/0402216. Bibcode:2004PhRvA..69f2320B.
4. Quantum Annealing in the Transverse Ising Model Tadashi Kadowaki, Hidetoshi Nishimori `https://arxiv.org/abs/cond-mat/9804280`
5. Kirkpatrick, S.; Gelatt Jr, C. D.; Vecchi, M. P. (1983). "Optimization by Simulated Annealing". Science. 220 (4598): 671–680. Bibcode:1983Sci...220..671K. CiteSeerX 10.1.1.123.7607. doi:10.1126/science.220.4598.671. JSTOR 1690046. PMID 17813860. S2CID 205939.
6. `https://arxiv.org/abs/1401.2910https://arxiv.org/abs/1304.4595`
7. "IBM Makes Quantum Computing Available on IBM Cloud to Accelerate Innovation". `https://www-03.ibm.com/press/us/en/pressrelease/49661.wss` 2016-05-04.

11

Qambrian Explosion

But to me, quantum computing is not an impossible dream; it is a possible dream. It is a dream that can be held without flouting the laws of physics as currently understood. It is a dream that can stimulate an enormously productive collaboration of experimenters and theorists seeking deep insights into the nature of decoherence. It is a dream that can be pursued by responsible scientists determined to explore, without prejudice, the potential of a fascinating and powerful new idea. It is a dream that could change the world. So let us dream.

—John Preskill

Seeing qubits in a cup of coffee

Making a real quantum computer available for anyone to try out on the Internet in May of 2016 will forever cement IBM's place in all future histories of the field. However, if the company hoped it would enable it to lock in the kind of dominance of this new technology it had enjoyed during the mainframe era, it would rapidly prove not to be the case. For many, the event served as a starter's pistol, creating more awareness than anything that had happened before, and waking the private sector to the technology's potential. Like the Cambrian Explosion 540 million years ago, when the fossil record shows a sudden, dramatic burst of biological diversity, quantum computing experienced its own proliferation — competing qubit modalities, startup companies, and national programs erupting seemingly all at once.

Quantum computing research had been experiencing diminishing interest from the public sector funding sources like DARPA. Since the original jolt of Shor's algorithm in 1995, the realization was dawning on the intelligence community that "breaking encryption" was going to take thousands and thousands of high-quality qubits, while all current efforts were topping out under 10, and low fidelity ones at that. Alternative approaches like D-Wave's annealers possessed more "qubits" but there was real debate regarding the technology's claim to quantum behavior or any possible advantage over classical computation, and the excitement about its launch was also dimming by the mid-2010s.

However, if the likes of IBM, Google and Microsoft were investing in internal R&D efforts, why shouldn't the private sector venture capital class take its shot? Perhaps sensing that Google and IBM's efforts in the field would make it easier to convince venture capitalists to invest, Chad Rigetti had left IBM to found the second quantum startup in existence, the eponymous "Rigetti," in 2013. In March of 2016, weeks before the IBM Quantum Experience launch, the startup raised $24M in an A round led by Andreesen Horowitz, with the intent to build superconducting qubits of their own.

Over in the ion trapping community, Chris Monroe was energized. He had left NIST in 2000, joining the University of Michigan to develop a highly capable group working in ion trapping, aiming to build more expertise and momentum for the technology. From there he moved to University of Maryland and the Joint Quantum Institute (JQI) where UMD collaborated with NIST in 2007. At the JQI, his group had demonstrated entanglement of two ions in separate traps using photons, a critical challenge for reaching commercial scale. He was ready to take the entrepreneurial plunge.

Monroe co-founded IonQ in 2015 with Jung Sang Kim with a $2 million seed investment from New Enterprise Associates, a VC from the valley. Kim was a researcher from Bell Labs with whom he'd been collaborating on the photonic interconnects. Kim had been working on optical switches for telecommunications networks at Bell Labs, a project that became less important as multiplexing on single fibers proved more efficient and effective. IonQ raised $20 million in 2016 in a round led by Google Ventures and NEA. Not only was IBM not without competitors in superconducting qubits, now the modality itself had a rival for the race to commercially viable quantum computing business. IonQ wasn't even the only ion trap project for long - the pioneers of ion trapping at Innsbruck University, including Zoller, Cirac and Blatt founded a spin off startup called Alpine Quantum Technologies and longtime defense contractor in the US Honeywell began their own ion trap qubit R&D effort as well.

John Martinis announcing the supremacy experiment on stage at a conference in 2019 was clearly Google's attempt to create the perception that the company had pulled away from the pack, but the reality was quite different. In addition to sharpening the rivalry with IBM, Google's move may have also played a role in announcements of cloud service offerings from Microsoft and Amazon Web Services in 2020. Both vendors planned to launch services that would provide access to multiple devices running different modalities of qubits. This seemed to be an explicit recognition that the market was not converging on a single vendor's qubits any time soon. And also, that Microsoft's hunt for Majorana particles, vital to their topological qubit design, was still not producing any promising results, so they would provide their customers with access to vendors whose efforts had proved more successful.

Within the year, both vendors had followed through with their plans. Microsoft's Azure cloud division launched Azure Quantum with access to IonQ's trapped ion device and the superconducting system from Rigetti. Amazon Web Services' Braket launched with IonQ and Rigetti as well as D-Wave's annealing system. Now customers wondering about quantum computing had a number of choices available to them — standalone web sites that offered access to a single vendor's device like IBM's, or multi-device services in the cloud with unified interfaces like Amazon Braket and Azure Quantum. That both services focused on superconducting and trapped ion systems from IonQ and Rigetti reinforced the market perception that the race to commercial viability lay between those two modalities, and soon the superior technology from the best vendor would soon deliver value. However, by the early 2020s no-one's qubits, superconducting or trapped ions, were capable of commercial, practical applications, only interesting experiments.

In fact, by 2023, it seemed as though perhaps all the investment and excitement in superconducting qubits may have been premature. Martinis had left Google in 2020, shortly after the supremacy experiment, and the company's chip R&D efforts seemed to have stalled, or at least had withdrawn from the public stage. Their cloud access for developers shut down in 2022, and while their group of world-class researchers produced a steady flow of top-notch papers, there wasn't much talk about bringing a product to market. Rigetti had been joined by several other superconducting startups, such as Oxford Quantum Circuits, IQM, and SeeQC, but no system had demonstrated any performance advantage over any classical computer for any task, and Rigetti's stock price had sunk nearly to the penny stock range. Rigetti himself departed the company in 2023.

IBM's resolve remained strong in 2023, at least in terms of product marketing. They had updated their product roadmap a few times since 2020, with the 2023 plan promising a 200 million dollar "hybrid quantum/classical supercomputer" with 100,000 qubits by 2030. Their public plans involved a data center sized collection of dilution refrigerators with as many qubits in each as they can fit, and classical links allowing parallel quantum computing at scale. Questions remained about IBM's program, however, namely how their chip topology can deliver greater connectivity between qubits while maintaining control over the system's noise. Connectivity is crucial because if two qubits are not physically connected on the chip, one can only entangle them through a series of operations called SWAP gates, which literally swap the state from one qubit to another. If this game of telephone between two qubits goes on too long the computation collapses into noise, the short coherence times on superconducting devices insufficient to complete any computation. Additionally, classical technology used to link quantum devices will impose significant computational overheads, which make it difficult to imagine realizing any true quantum advantage.

Meanwhile, Microsoft's topological qubit program was in shambles. In 2022 Kouvenhoven had to retract a paper his group had written that claimed to find signatures of Majorana particles in a device, due to mischaracterization of the results[1]. The mistake may have been unintentional but was likely a reflection of the mounting pressure on the researcher to deliver progress for his corporate source of funding. Charlie Marcus also had a published result challenged, though in his case the independent investigation found nothing amiss[2], it was merely a result of being in the "blast zone." Both men left Microsoft, and the company's topological qubit program showed no signs of recovering from their departure.

The uncertain fates of privately-funded superconducting qubits over the decade since they were launched is a cautionary tale about the nature of the challenges still faced by the technology. Josephson junctions may have provided a macroscopic stage on which quantum effects could be observed, but controlling those phenomena is by no means a solved problem. At this stage, the best minds in the world can fabricate a superconducting chip with little insulating gaps at just the right spots, can hook that chip up to microwave cabling in the bottom of a dilution refrigerator, cool the whole thing to eleven thousandths of a degree Kelvin above absolute zero, and then use room temperature electronics to talk to that chip, and every aspect of that is a minor miracle. However, the road to fault tolerant quantum computing at large scale is blocked by knowledge we simply do not have today. Without large numbers of qubits able to precisely execute long-running programs, we have very interesting lab instruments, but not industrial-scale computers. Every so often a claim will be made that the science of superconducting qubits is "solved," and all that is left is incremental improvements in the systems engineering, but that is simply not true. The artificial atom may eventually have its day, or it may, like the Nuclear Magnetic Resonance qubit, terminate its forward progress, a fascinating bit of science with no practical use, abandoned and eventually forgotten.

Things in the ion trap camp were not faring much better. IonQ had been joined by another commercial ion trap device vendor, Quantinuum, which spun out of Honeywell's quantum computing R&D program in 2021[3]. Both had very capable systems with about 30 qubits each, and both faced the same limitations of their approach — those of scale and speed. It was certainly true that trapped ions had much, much longer coherence times than superconducting qubits, however, gate operations were so time consuming that the trapped ion's effective "clock speed," or rate at which it could carry out operations, was a couple of orders of magnitude slower than superconducting systems, the equivalent of a classical computer running at a few kilohertz. Given that modern classical computers run in the billions or even trillions of cycles per second, that performance gap seems important.

More troubling is the question of scaling up ion trap-based quantum computers. One dimensional traps, where the ions are held in a single line, can only contain around 30 qubits before the thermal energy generated by gate operations overwhelms the capacity of the lasers to maintain the required temperature. A number of schemes for scaling up are being explored. Some traps consist of two one-dimensional traps in an "X" formation, and the qubits from one trap can interact with those from the other. The lasers used to cool and apply gate operations are also held up as a possible means to optically link traps, however this entanglement operation is painfully slow, as the photons only interact with the ions probabilistically, failing as often as it succeeds. Alternative trap designs are being experimented with as well. Oxford Ionics has a Paul trap implemented on a chip that also has an embedded antenna, which allows them to apply gates through microwaves, similar to superconducting devices, which promise faster operation and larger scale chips[4]. Universal Quantum plans to solve the problem of linking qubits across traps by having the individual ions actually *leap* from one trap to another.[5] Others are returning to research Penning traps, hoping to build two-dimensional arrays of micro traps, each ion in its own magnetic field, another theoretical solution to scaling.

Like their colleagues working with superconducting devices, the ion trappers have achieved incredible feats of engineering and creativity with their devices. Charged atoms are produced and deposited into a trap with automated efficiency and precision. Coherence times started off impressive and have continued to improve. Gate fidelities are the best of all existing qubit types. The devices are miraculous achievements, and ion trap-based clocks have achieved a stunning precision of 9.4×10^{-19}, meaning that, if one could run for 33 billion years, twice the age of the known universe, it would still be within one second of accuracy.[6] However, they are still not practical quantum computers capable of carrying out tasks faster than your average laptop circa 2023. Actually, most things a laptop can do an ion trap device can't even begin to replicate. In 2023 Chris Monroe left IonQ[7], and Tony Uttley, who had led the effort at Honeywell and been Quantinuum's CEO at the spin out, also left[8]. IonQ's official statement about Monroe's departure was that the fundamental science was done, and Monroe wanted to get back to academia since the company only had engineering challenges ahead of it. A charitable interpretation would be that it is difficult for a world-class physicist to lead engineering efforts, but perhaps it was that the challenges of commercial qubits at this point were more of the entrepreneurial variety.

The narrative of the two-horse race between superconducting Josephson junctions and trapped ions as qubit modalities was weakening as well. Both approaches and all the variations being explored of each are so far from being solved problems that the field is effectively wide open. A number of alternative qubit modalities have emerged, some with roots going as far back as ions and superconducting technologies, some with more recent origins. Several are currently being researched, and even though most are still very early stage, others are already approaching the maturity and capability of either atom- or superconducting circuit-based systems.

It is sometimes said that both trapped ion and superconducting qubits are using electrons for computation. In superconducting qubits, of course, the current in the circuit is made of electrons, which form the artificial atom that provides the basis for the qubit. Trapped ion qubits, while based on the entire atom, use lasers to excite the electrons of that atom to different states to represent the |0> and |1> states. Both these qubits' vulnerability to noise is in large part due to their reliance on electrons, because their state is very easily disrupted by heat, by electromagnetic interference, even by cosmic rays.

There is a subatomic particle available in abundance, much less sensitive to noise from heat or other sources, that also can serve as a quantum two-level system suitable for computation — the photon. Theoretical evidence for the photon's existence was provided by no less than Albert Einstein himself, in a 1905 paper explaining how the photovoltaic effect works[9]. Initially Einstein used the term "quanta" to describe the discrete units in which light occurred, a usage that helped pave the way to the general adoption of both the phrase and the theory of quantum mechanics. An American physical chemist, Gilbert Lewis, popularized the term "photon" in 1926, adapted from the Ancient Greek, *photos*, for light.[10] Of course, photons had been inextricably linked to the developments leading up to quantum mechanics, all the way back to Young's double slit experiment, which, though Young didn't know it at the time, used photons to demonstrate superposition.

The roots of quantum information science are similarly associated with photons. Stephen Wiesner's breakthrough thought experiment from the late 60s, published over a decade later[11], used polarized photons to represent the serial numbers of his money that would be impossible to counterfeit. Around the same time, in 1972, quantum entanglement was first demonstrated using photons. Charlie Bennett and John Smolin experimental cryptographic key exchange was inspired by Wiesner's work, and used polarization in much the way he had suggested.[12]

Despite these milestones and ongoing use as a workhorse in physics labs, where optical tables, lasers and pieces of glass are habitually used in a broad set of experiments, photonic quantum computing did not develop at the same pace as trapped ion or superconducting qubits. Progress, however, has been made, and several companies are working on photonic qubits, grappling with their own set of hurdles to get to quantum advantage.

One notable example is PsiQuantum, founded in 2015 by Jeremy O'Brien, Terry Rudolph, Mark Thompson and Peter Shadbolt, physicists working at several UK universities. O'Brien had experimentally implemented a CNOT gate on photons in 2003[13], and Rudolph contributed a unique approach to photonic-based computation the firm refers to as "fusion based quantum computing." Psiquantum had raised almost $800 million by 2023 but its progress is difficult to gauge as the company has famously chosen to remain in "stealth mode" until it can demonstrate fault-tolerant computing with a large number of qubits.

Other photonic companies have been less secretive, such as Xanadu Quantum Technologies, founded in Toronto in 2016, out of the Creative Destruction Labs' quantum track. Xanadu's approach initially used squeezed states of photons to perform Gaussian Bosonic sampling, which promised better performance for a certain subset of problems. In 2022 the Xanadu team demonstrated a device they called Borealis to perform computations they claimed were 50 million times faster than the classical alternative.[14] Like Google's supremacy experiment, the problem was contrived as a demonstration and had little practical value, and the Borealis device was not a universal computer. As a non-gate based device, Borealis was something of a revival of approaches like quantum annealing. While analog quantum annealing devices like D-Wave's had not shown any practical advantage in the 15 years since founding, others were still considering an analog approach.

While the clock builders at NIST had been focused on charged atoms in their ion traps, there existed a larger segment of the atomic physics community who had been working with neutral atoms. Before arriving at NIST, Chris Monroe had been building neutral atom devices at the University of Colorado in Boulder before joining NIST, and it was at Boulder where some of the most important work on neutral atoms were performed. The motivator for these experimental devices was a state of matter theorized by work by Satyendra Nath Bose and Albert Einstein in the 1920s. At extremely low temperatures, atoms would share the same quantum state, effectively demonstrating quantum behavior at macroscopic scale. In 1995, the Bose-Einstein condensate (BEC) was realized in the lab at Boulder by Eric Cornell and Carl Wieman, earning them a Nobel Prize in 2001.

While the pursuit of Bose-Einstein condensates and the Nobel prize everyone assumed would go with it was on, neutral atoms attracted more physicists and funding than trapping ions. Once BECs had been created, that community began to think of other applications for their expertise, and as quantum computation became more imaginable, their attention naturally fell on that topic. Lacking a charge to manipulate via electromagnetic fields, neutral atoms must be held in an optical lattice created by lasers. Photons are also used to cool and then to excite the atoms, creating what's called a "Rydberg state" so that the outer electron orbitals can be made to interact with one another.

The Rydberg state of an atom is difficult to manipulate, as it is highly excited and therefore prone to noise, however, Mikhail Lukin, working at Harvard, was successful in building a stable array of 256 atoms in the summer of 2021. The atoms in this device are sometimes referred to as qubits, however, the term is misleading, because Lukin's device did not support gate operations. Instead, the atoms' energy and position could be adjusted and tuned to model the time evolution of a specific Hamiltonian equation.

A Hamiltonian, named after Sir William Rowen Hamilton, who created it in 1833, is a crucial equation for understanding isolated dynamical systems. Loosely speaking, it is used to describe the total energy of a system, both kinetic and potential, and is used with Schodinger's equation to determine how the quantum state of a system changes over time. Energy levels are used to predict the outcomes of interactions and transitions within the system, and so an efficient method for modeling a Hamiltonian is essential to the challenge of quantum simulation. In this way, what Lukin's group had built at Harvard was a realization of Feynman's original vision in 1981, an efficient means of simulating quantum systems, built using quantum mechanics.

Progress on the machine at Harvard had been promising enough for Lukin to found a company, QuEra, along with colleagues Vladan Vuletić, Markus Greiner, Dirk Englund, Nathan Gemelke, and John Pena in 2018[15], and in 2021, QuEra made its first commercial device, Aquila, based on the prototype in Lukin's lab, available on the cloud via Amazon's Braket service. Other companies are also pursuing neutral atoms, such as Pasqal in France, and Atom Computing and Infleqtion, both based in Colorado. Infleqtion was formerly known as Cold Quanta, and actually predates the others by more than a decade, founded in 2007, focusing initially on BEC before pivoting to computation more recently.

It turns out there are many ways to make qubits in addition to trapped ions, or superconducting circuits - not just photons or neutral atoms, either. While working towards their experimental NMR qubits, Isaac Chuang and Neil Gershenfield wrote in Scientific American that a quantum computer "might resemble [a] cup of coffee."[16] Two level systems abound in the quantum realm, any of which might be fodder for a novel qubit design. Companies like Quantum Brilliance are using defects in diamonds where nitrogen atoms take the place missing carbon atoms as qubits - known as nitrogen vacancy, or "NV" qubits. EeroQ, a Chicago-based company, is trapping single electrons on the surface of superfluid liquid helium, and using those as qubits. In many ways, however, the race to qubits can be reduced to a question of electrons versus photons. Even in devices that use entire atoms in radio frequency or optical traps, it is the electrons' energy state or spin that is used to encode information.

Using electrons feels familiar, of course, since electrons flowing through circuits are the basis for classical computing. In some ways the dream is to replicate the scale accomplished by the titans of classical, leveraging CMOS technologies to build quantum computers of the future. The most direct attempts to use CMOS are called spin qubits, where electrons are trapped in tiny wells or dots on the surface of an otherwise very familiar semiconductor chip. Spin qubits were first proposed by David Divencenzo and Daniel Loss in 1997[17], and in April 2020, QuTech at Delft University made a system with two spin qubits available to the public [18]. Spin qubits use the intrinsic 1/2 spin in the degrees of freedom of individual electrons, with the benefit of being fabricated using our familiar CMOS technologies. Like photonic qubits, they promise massive scalability and more practical deployment, if they can overcome certain critical challenges to implementation. Several startups have been founded to work on silicon-based spin qubits, including Diraq and Silicon Quantum Computing in Australia, Equal1 in Ireland, and Quantum Motion in the UK. In addition to startups, Intel has made a serious investment in spin qubits, leveraging its unparalleled experience and resources. In 2023 they announced Tunnel Falls, a 12-qubit silicon chip, available for researchers. [19]

With CMOS-based spin qubits a possibility, one might ask why startups like QuEra, PsiQuantum and IonQ and even giants like IBM are bothering to pursue their exotic qubits modalities. Like each of their approaches, spin qubits on silicon face multiple challenges that require fundamental scientific breakthroughs. In basic research there really is no way to gauge timelines or even feasibility, a scientific Black Swan with no way to measure the odd of success or failure. More exotic ideas about how to build qubits continue to crop up around the world, all of which face numerous unsolved scientific challenges before they can be usable. This evolutionary diversity only serves to emphasize that in the 2020s, forty years past the nominal start of the race, there simply is no low-risk path to making devices large enough, stable enough and performant enough to win over all the others. Or to do anything useful to industry, for that matter.

What one is facing is a gaping hole in our knowledge, and no proof that void even has a viable solution to fill it. In typical industrial research and development, the focus is more on engineering challenges which are incremental in nature - issues of miniaturization or finding the right material or making fabrication more accurate or cheaper. Each team working on qubits is making their own (highly) educated guess about the solvability of their modality's unique challenges. Ion trappers believe in the fundamental stability of their qubits and believe gate speed and scalability won't prove insurmountable. The superconducting community believe their fabrication skills will continue to improve and their systems' speed is unmatched, and don't worry about the short coherence times, connectivity challenges or their reliance on dilution refrigerators and Helium-3, an isotope of Helium that is only produced as a by-product of radioactive decay. The photonic proponents point out that photons make the most stable qubits and maintain that they can overcome the difficulties of interacting with them or losing them to absorption, and anyway scaling any of those other types of qubits will require using photons as "flying qubits" anyway.

Each variation on a modality of qubit, whether from a VC-backed startup or a corporate-backed R&D team, arcs out into this wash of unknown unknowns, a flare describing a path up into the darkness, its apex hopefully finding commercial usefulness. The truth is, no existing qubit modality has a clearly defined path to being stable enough at a large enough scale to be anything more than very elaborate academic experimental devices. With all the investment in a technology that has not yet proven itself to be valuable, it's no wonder that some are reminded of the dot-com bubble and other instances where the hype has led the public astray.

References

1. Zhang, H., Liu, CX., Gazibegovic, S. et al. Retraction Note: Quantized Majorana conductance. Nature 591, E30 (2021). https://doi.org/10.1038/s41586-021-03373-x
2. Guéron, Sophie, Pertti Hakonen, and Allan MacDonald. "Science Paper "flux-induced Topological Superconductivity in Full-shell Nanowires", Report from Expert Panel". Zenodo, February 15, 2024. https://doi.org/10.5281/zenodo.10647080.
3. "Introducing Quantinuum: The World's Largest Integrated Quantum Computing Company". finance.yahoo.com. November 30, 2021. https://finance.yahoo.com/news/introducing-quantinuum-worlds-largest-integrated-220000928.html
4. Malinowski, M., D. T. C. Allcock, and C. J. Ballance. 2023. "How to Wire a 1000-Qubit Trapped Ion Quantum Computer." arXiv. https://doi.org/10.48550/ARXIV.2305.12773.
5. Valahu, Christophe H., Iason Apostolatos, Sebastian Weidt, and Winfried K. Hensinger. 2022. "Quantum Control Methods for Robust Entanglement of Trapped Ions." arXiv. https://doi.org/10.48550/ARXIV.2206.06064.
6. https://www.optica-opn.org/home/newsroom/2019/july/optical_clock_precision_breaks_new_ground/
7. https://ionq.com/news/ionq-announces-senior-leadership-transition
8. https://thequantuminsider.com/2023/11/16/tony-uttley-to-step-away-from-quantinuum/
9. Einstein, A. (1905). Über einen die Erzeugung und Verwandlung des Lichtes betreffenden heuristischen Gesichtspunkt [On a heuristic point of view concerning the production and transformation of light]. Annalen der Physik, 322(6), 132-148. DOI: 10.1002/andp.19053220607
10. "December 18, 1926: Gilbert Lewis coins "photon" in letter to Nature". www.aps.org. Retrieved 2019-03-09.
11. S.J. Wiesner, "Conjugate Coding", SIGACT News 15:1, pp. 78–88, 1983.
12. Smolin, J. A. (2004). The early days of experimental quantum cryptography. In IBM Journal of Research and Development (Vol. 48, Issue 1, pp. 47–52). IBM. https://doi.org/10.1147/rd.481.0047

13. O'Brien, JL; Pryde, GJ; White, AG; Ralph, TC; Branning, D (2003). "Demonstration of an all-optical quantum controlled-NOT gate". Nature. 426 (6964): 264–267. arXiv:quant-ph/0403062.
14. Madsen, Lars S.; Laudenbach, Fabian; Askarani, Mohsen Falamarzi; Rortais, Fabien; Vincent, Trevor; Bulmer, Jacob F. F.; Miatto, Filippo M.; Neuhaus, Leonhard; Helt, Lukas G.; Collins, Matthew J.; Lita, Adriana E. (1 June 2022). "Quantum computational advantage with a programmable photonic processor". Nature. 606 (7912): 75–81.
15. Russell, John (23 November 2022). "QuEra's Quest: Build a Flexible Neutral Atom-based Quantum Computer". HPCWire. Retrieved 18 July 2023.
16. https://www.thekurzweillibrary.com/quantum-computing-with-molecules
17. Loss, Daniel, and David P. DiVincenzo. 1998. "Quantum Computation with Quantum Dots." Physical Review A. American Physical Society (APS).
18. `https://qutech.nl/wp-content/uploads/2020/04/2.-Technical-Fact-Sheet-Quantum-Inspire-Spin-2.pdf`
19. `https://www.intel.com/content/www/us/en/newsroom/news/quantum-computing-chip-to-advance-research.html`

Subscribe to Deep Engineering

12

The Heroic Era Of Qubits

> *The man who is successfully to invent a new paradigm must necessarily be insulated from the normal science... he must have a certain blindness to the existing rules. [...] But normal science does not aim at novelties of fact or theory and, when successful, finds none. The new paradigm forces scientists to investigate a new world, [and the revolutionary] is superseded by the mop-up work of normal science.*
>
> — *Thomas Kuhn*

Big, hairy, audacious goals

From the start, the scientists who began speculating about reformulating information theory according to the axioms of quantum mechanics were viewed with skepticism. On the one hand, the digital age was doing just fine within the constraints of "classical" information, and on the other, quantum mechanics is so strangely exotic, so outside of human experience and intuition, what would a "quantum information science" even mean? Still, the irresistible force of human curiosity persisted, and gathered adherents and new champions, and slowly, year by year, built the case for quantum information science and computing. After Endicott in 1981, a small group began an annual conference in Torino, Italy, where the emerging ideas and the theoretical progress could be shared. As he had at Endicott, Charlie Bennett would take a group photo each year, and each year he had to step back a bit further to get all the attendees in the frame. The participants weren't just keeping the lights on, they were gathering new converts.

By the early 90s, Wheeler was asking his undergrad students at UT Austin if they wanted to take a stab at recapitulating the axioms of quantum mechanics in purely informational terms. At some point David Deutsch and Richard Jozsa visited Wheeler, along with Feynman, Bennett, and the rest of the true believers. "It from bit," Wheeler's most famous of his pithy phrases, captures his growing belief that the universe itself was a kind of computer, and so building a device that used its fundamental rules to manipulate information just made sense. After

Deutsch, the theory seemed solid. After Shor, incontrovertible. The focus of skepticism shifted to the impossibility of building a real device that could realize the theoretical advantages.

Since the late nineties the field has seen tremendous advances. with coherence times increasing in leaps and bounds and noise levels slowly falling. These advances have been hard won by the experimentalists and engineers working their way along paths suggested by theorists, by material scientists, by veterans of CMOS fabrication, algorithmists, by everyone who has cared enough to invest their time thinking deeply and working hard on this enormous challenge. Mid-way through the 2020s, we can look back over the past decades and marvel at how far the field has come. Physical devices with hundreds, even a thousand qubits are being made, which seemed like an impossible dream just twenty years earlier when the research efforts seemed stuck below ten qubits. Real demonstrations of fault tolerant "logical" qubits have occurred on trapped ions, superconducting qubits and neutral atom arrays, suggesting that very long running, precision tasks impractical for a classical computer or a NISQ device might soon be possible.

What we don't have, however, is a clear view on just how long the vision of thousands of fault tolerant qubits capable of large scale, commercial workloads might take. Or if it's even possible. Each existing qubit modality has its own fundamental challenges to overcome in order to achieve scale and stability, and solutions to those challenges are far from clear. Classical computing keeps progressing as well, with GPUs accelerating increasingly sophisticated machine learning applications and accomplishing feats previously thought impossible, like beating the best human at chess, or Go, or predicting protein folding. Even classical information theory has raised the bar on quantum computing, meeting advances in quantum algorithms with its own advances that deliver efficiency gains without having to rely on the existence of anything more exotic than our current microprocessors.

Against that backdrop, it's clear that the biggest risk faced by the field is a loss of faith in its potential. Almost simultaneously with the private sector funding surge of the 2010s came the talk of a quantum winter. Based on the rollercoaster of hopes and disillusionment the field of "artificial intelligence" has been riding since the 1970s, observers worried that overly optimistic expectations for quantum computing progress would collapse and jeopardize what progress was being made. So far that hasn't happened, precisely, but certainly expectations have had to be tempered and the flush of excitement that followed the Supremacy experiment has diminished and shifted, largely, back to classical machine learning and the large language models from OpenAI and others.

At this point it's time we took stock of where the field is the development of usable quantum technologies. In the mid 2020s, more than forty years since the Endicott conference, thirty years since the factoring algorithm, more than twenty since the first working physical qubits, more than five years since the Supremacy experiment, where are the commercial applications of quantum computing? When Shor unlocked a torrent of anxious government spending hoping to fund a code breaking machine before their adversaries, it also awoke a desire among the physics community to lead that effort, delivering qubits of sufficient fidelity at sufficient scale to achieve the miraculous, and possibly earn a Nobel doing it. This should not be surprising. The Nobel Prize is the ultimate recognition of a physicist's abilities, and for many, there is no higher motivation. The path towards a practical quantum computer already has produced a number of Nobels, including Josephson's startling discovery of electron tunneling when he was just 22, now the basis of superconducting qubits, Wineland's prize for trapping ions, and most recently, Clarke, Devoret, and Martinis' work building on Josephson's breathrough. Once the first working devices were created it was reasonable to assume that a physicist would birth fault tolerant qubits into the world like Zeus issuing Athena forth from his forehead, and so collect a Nobel for his efforts. Also reasonable was the assumption it would be Zeus giving birth, not Hera, due to the very patriarchal and often outright misogynistic world of physics. Add to the mix the reputation for arrogance among physicists and it seems inevitable that a cadre of men would become convinced of their ability to change the world through quantum computing in the early 2000s. As of 2026, no such mythic act of creation has occurred, with no corresponding Nobel.

This is in no way to disparage those scientists, diminish their skills or underestimate the enormity of the challenges their personal odysseys would lay before them. That group, which includes many people already introduced in earlier chapters, such as Chris Monroe, Jay Gambetta, Chad Rigetti, Jeremy O'Brian or John Martinis, looked at the earliest implementations of qubits and saw them as tantalizingly close to something useful for uses beyond exotic lab instruments. Once the the intelligence community convinced itself the *Sneakers*-like crypto-apocalypse was far enough in the future to ignore for now, the less commercially-oriented dollars became much harder to come by, and those early pioneers struggled to keep their teams on track. Some began raising capital from venture funds, from public sources of funding aimed at technology commercialization, or from corporate R&D budgets. They embarked on projects with five to ten years, sometimes more of operating budget secured. They began to behave and to look like tech startups.

Tech startups are, at this point in the digital age, a very, very familiar species to us. The attributes of the earliest stages of startups are part of the cultural fabric of the 21st century. There were those who, seeing a quantum startup, may have found it easy to evaluate that firm and its activities using the familiar trope of the tech startup, and to make assumptions about its likely future accomplishments. Some of the physicists who founded these quantum startups may have struggled a bit with this same delusion. Or perhaps some were clear-eyed, but did not go to great lengths to disabuse their investors if they heard them use phrases that indicated category confusion.

If this book has managed to convey one fact thus far, we hope it is this: quantum computing is not in the same category as any emerging technology within modern memory. It isn't, in fact, just a new category, it's an actual new paradigm of computing, honest-to-goodness, not like the other million times you heard that word used hyperbolically. We've seen that the ways in which you build a qubit are not the ways the pioneers of classical computing built bits. We've seen that the math that is the native language of qubits is not native to classical bits, and so the algorithms look *very* different. The nature of a paradigm shift is that it is discontinuous, and nothing that exists before it will accurately predict the world that exists after the shift occurs. This is in stark contrast to the incremental progress classical computers have enjoyed since their last true paradigm shift, the invention of the microprocessor. The engineers hand-soldering transistors to boards to build guidance computers for Apollo spacecraft could not have imagined chips infiltrating our kitchen appliances, children's toys, cars and Hollywood movies. Put another way, from a contemporaneous perspective, the launch of the iPhone was impressive but many companies had tried their own versions of that device already, and we all had a sense of what the mobile Internet would bring us, at least in broad strokes.

That cognitive framing is also deeply embedded in the funding models dominant in the private sector for funding new and emerging technology development. The typical venture fund has a horizon, past which the venture capitalists or VCs managing the fund need to start exiting their investments so that they can show returns for investors who are the sources of the capital in the fund. Corporate CFOs similarly have time scales on which they need to show returns on technology R&D spending in the form of new lines of business growing revenues. The time scales vary somewhat from shorter to longer, but very rarely do they exceed 10 years, which is comfortable when incrementally improving on existing technology but can be problematic for more fundamental scientific research. This has led to the decline of the great industrial R&D labs of the latter half of the 20th century like Xerox Parc or Bell Labs, the engines of so many massive leaps forward in computation and communication.

So these great men seeking to build practical qubits had little choice but to find compelling arguments for the CFOs and the VCs to make a bet on their grand projects. Certainly, all of their pitch decks held warnings that this was not your dad's iPhone or dating app you were investing in, it was a longer term and more uncertain bet than that, but these founders were passionately excited about what they wanted to build, and that came across loud and clear. Additionally the founders had the advantage of speaking in very esoteric jargon about very complex and profound exercises involving our best understanding of the nature of the universe to the VCs, who as a class never like to know less than anyone else in the room, and as a defensive maneuver will just nod sagely instead of admitting when they are lost.

Not that these were dumb bets, not at all — a new paradigm for computing has only happened once before in modern history and the earliest investors in the successful applications of that paradigm made quite a bit of money. The ones that timed things correctly did, in any case. The timing is crucial, of course. Companies that existed before Fairchild Semiconductor and Texas Instruments are mostly forgotten, other than IBM who may be alone in the ability to claim they watched the technology mature and chose the opportune moment to invest. Or maybe they saw their tabulation machine business at grave risk of disruption at the hands of digital computers and made a savvy leap of faith.

The issue with timing is that to get it right through rational planning one must have a way to model out when the relevant events will occur. This is why smart money hates investing in fundamental science. Scientific breakthroughs are almost always achieved by human curiosity and undirected inquiry - the only two strategies proven to be effective when we "don't know what we don't know." These scenarios are what Nassim Nicholas Taleb calls "black swans," a concept he developed beginning in his 2001 book *Fooled by Randomness* and fully articulated in 2007's *The Black Swan*. The term derives from the belief that all swans were white until a black one was seen in Australia. Black swans are challenging for investment not only due to uncertain timelines but more fundamental inability to measure the chances of either a positive or negative outcome. Like the swans, before you see the black one, 100% of swans are white, and the chance of there being a black swan is 0%. After the discovery, the probability of there being a black swan has increased infinitely. This is a difficult set of circumstances to construct a rational investment case around.

In the early 2000s, no qubits were good for anything of commercial interest. The likelihood of a qubit existing in the future that would have commercial value was entirely unknown, and in fact had a non-zero probability of being a complete impossibility. Into that monumental landscape of uncertainty entered a cadre of physicists who each believed in his own ability to find a path to perfect qubits in sufficient numbers to be commercially valuable. Like Alexander the Great storming enemy cities at the very front of his armies, these men leapt into the fray, assembling teams of talented scientists, distilling speculative R&D roadmaps into pitch decks and convincing sources of capital to invest, all based on uncut self-confidence and little else. This is not unusual behavior for physicists. An ambitious physicist must constantly take action in the face of complete lack of foreknowledge, and only the example of incredible advances in human understanding originating from the intuition and hubris of the physicists who came before. As physics has advanced, we have gained truly incredible insights into the way the entire universe operates, shedding light on every way in which we exist within it. It is sometimes said that physicists believe that all other sciences can be derived from physics. It is probably said in jest.

These physicists' self-confidence served them with investors, as well. Of course, technology investment very often includes some optimism and even over estimating one's ability to achieve milestones. "Fake it 'til you make it" is the operative phrase in Silicon Valley. That the subject was incredibly esoteric and these investors often found themselves entirely dependent on those same physicists to help them understand the pitch decks and milestones added to the risk of an echo chamber. Each modality of qubit ended up with a small number of true believers, little groups of physicists, some technologists, some entrepreneurs, each perfecting their pitch and looking for their investors. The best pitches balanced a certain amount of irrational exuberance with caution about uncertainty, found their money in both venture funds and corporate coffers, and got down to doing the hard work.

It is undeniable that their efforts have been producing impressive steps forward. We've evolved far beyond the tiny handfuls of short-lived qubits to near maximum yields for this stage in the evolution of each modality. Superconducting devices from IBM have broken records, exceeding 1000 qubits on a single chip. Ion traps are at the limit for one dimension, above 30 ions in stable, long lived coherent chains in the machines from IonQ and Quantinuum, with two-dimensional prototypes in the labs of Universal Quantum, Oxford Ionics, university labs and elsewhere. Ion traps' cousins, neutral atom arrays, already in 2-D lattice arrangements, are developing rapidly in devices from QuEra and Pasqal, starting as analog simulator devices and now working on digital gates along with others in the US and Europe. Photonic and spin qubit boosters continue to ask for patience and promise startling future accomplishments.

The gate fidelities on all devices continue to increase, leading to a flurry of fault-tolerant demonstrations and announcements in 2023 and 2024 from IBM, QuEra, Quantinuum and others. And the pack of quantum hardware firms do not have anything like a lock on the state of the art in their chosen qubit modalities. In superconducting devices, other corporate and startup efforts everywhere from Amazon Web Services to France's Alice & Bob or Fujitsu in Japan continue to work on novel implementations which may leapfrog the limitations of IBM's or Google's architectures. Exotic, wild ideas for alternative qubits lurk in lesser-known labs. Most promising, the mid-2020s are starting to see physical science experiments on hardware producing results that would be difficult if not impossible to achieve by classical computational means. The Feynman machines are arriving.

The overarching challenge that unites the field is to find the balance required to stay the course. Enterprises who will eventually be the using quantum computers to achieve new operational capabilities must calibrate their investments to maintain a focus that is sufficient and sustainable. Sufficient to ready them for rapid adoption when the time is right, at a level that is sustainable over an unknowable span of time. Investors, too, must be rigorous with their own expected horizons on returns. Either settle in for the long haul, using funds that can roll from one ten-year fund to another if need be, or target shorter term, lower yield exits through supply chain plays, professional service models, acquihire and the like. The public sector also has a vital role to play. It is far too early to step aside for the private sector to take the lead on this highly strategic technology. Innovative private-public partnership models need to be experimented with on a continuous basis at the regional, national, and global scale, until such time as the arc of development surges past any nurturing framework to take its place as a spectacular source of the creation of value.

Most importantly, the scientists and engineers dreaming of our quantum future need to practice radical transparency on all existing caveats. Allowing narrative smudging arriving from gaps in understanding, or imprecise analogies with classical emerging technology patterns, or other errors may keep the funding train on the track in the near term but eventually the bills will come due. The crucial mission is to make the case for the transformative power of operating "at the quantum limit," as Dana Anderson, one of Cold Quanta's founders, calls it. Quantum physics remains mystifying even as it continues its history as the most successful scientific theory in history. The computing and communication technologies we are building derive their power directly from the most mystifying aspects of that theory, and therefore are highly resistant to the human imagination. In this way, a central challenge of quantum information science is one of communication. We desperately need the metaphorical and narrative tools to begin to grasp what it is we are building, to both maintain our focus on the task at hand and to prepare for the future these tools will enable us to create.

Part 3

What We Know and What We Don't

Relying on mainstream tech press or worse, marketing materials, can create any number of false impressions about just how soon we'll be using quantum computers in production or what they will be good for. We'll examine the reasons why so much confusion surrounds the technology, and how that confusion can be diminished but probably never completely dispelled. We will dive deep into the major blockers to a new quantum era, namely scale and fault tolerance, and learn why quantum error correction is so monumentally hard. Finally we'll survey the current landscape and imagine the road ahead, what do the next five years or so hold for this new field and these exciting technologies?

- *Chapter 13 - Transcendent Quantum Hype*
- *Chapter 14 - The Road to Fault Tolerance*
- *Chapter 15 - The Quantum Horizon*

13

Transcendent Quantum Hype

To the layperson, 'quantum' has become a synonym for 'magic.' It's invoked to explain everything from free will to consciousness to homeopathy. But quantum mechanics is not a blank check to write your own reality. It is simply the operating system that the universe runs on. And just like any operating system, it has extremely strict, unbreakable mathematical rules about what you can and cannot compute. It doesn't let you bypass the laws of physics; it just gives you a slightly weirder set of laws to obey.

— *Scott Aaronson*

In which the hippies save quantum physics

Irrational exuberance about a new technology is a familiar phenomenon, and bubbles of over-valuation and over-investment are common enough for talk of bubbles to enter the mainstream discourse. The dot-com era is largely to thank or blame for the phrase "tech bubble" emerging into the popular imagination, with talk of the stickiness of eyeballs and S-1 filings in advance of Initial Public Offerings admitting to business models that could never be profitable[1], but it was not the beginning nor the end of the tech hype cycle.

Every emerging technology from 5G to cryptocurrency has had a moment when its hype men have made the case for how the world was about to change irrevocably convincingly enough that the markets took it seriously, at least for a while. The pattern is so prevalent that the technology analysis firm Gartner has a branded methodology for locating a particular technology along that hype cycle path[2], starting from an innovation trigger, heading up to a peak of inflated expectations before crashing into a "trough of disillusion." Technologies with legitimate value climb back up over time on what Gartner calls the slope of enlightenment and eventually find the plateau of productivity.

In the case of quantum computing, there's no doubt that the hype cycle was triggered by Peter Shor's discovery of the factoring algorithm in 1994. Before that the topic was so esoteric as to be non-existent in the IT community, a hobby horse for eccentric scientists. Shor set off a lightning fast reaction of excitement about the algorithm's seemingly magical power to break encryption followed almost immediately by the first wave of skepticism — surely the problems of noise and error correction would prevent the algorithm from running at the requisite scale. Peter Shor reacted to the skepticism by creating the first scheme for quantum error correction.

The field has been resilient in that way ever since — every year the skeptics decry the possibility of the devices ever working well enough, and every year gate fidelities go up, qubit counts go up, demonstrations of error correction become more and more impressive. Still, the skeptics persist, and they likely will until a large enough fault-tolerant device is put into service. At this stage the concerns and critiques are still rooted in the physics foundations of the field, and generally amount to some objection about the impossibility of sufficiently mastering the problem of noise.

Scott Aaronson, a theoretical computer scientist at the University of Texas who played a role in Google's Supremacy experiment, collected the arguments he was hearing most consistently and reviewed them in a lecture, the notes of which make for an entertaining read. Aaronson rebuts each effectively, finding no credibility in the claims that it can't possibly work, and concludes, "...it's entirely conceivable that quantum computing is impossible for some fundamental reason. If so, then that's by far the most exciting thing that could happen for us. That would be much more interesting than if quantum computing were possible, because it changes our understanding of physics."[3]

Of course, this isn't the first controversy sparked by quantum mechanics - recall the tensions that were present during the Solvay conference and the entire formative stages of the field. Initial skepticism that the universe existed in discrete quanta fell away as the experiments validated the theory again and again. Quantum understanding provided the underpinning not only for science but also new technologies, especially as the world moved towards the Second World War, culminating, of course, in the invention of the atomic bomb. In the post-war era, lasers and semiconductors continued to provide undeniable proof that quantum stuff worked. The problem that had so troubled Einstein and formed the basis of the EPR paradox — entanglement — was put aside in favor of the practical applications made possible by the rest of quantum mechanics.

This was the *status quo* in the years following WWII, but in the aftermath of the Manhattan project, the scientific community was experiencing deep uneasiness regarding the applications of their work by the military-industrial complex. This malaise was especially pronounced in younger scientists and students studying science, and so it was that the less practical aspects of quantum theory resurfaced in the halls of universities the world over.

Entanglement in particular provided a topic of inquiry that not only seemed unlikely to lead towards novel military applications but also felt aligned with some of the ideas percolating through sixties counter culture from eastern philosophy, mysticism, and other challenges to the hegemony of the western mind. Thus it was that, in 1975 the Fundamental Fysiks Group was formed by grad students attending University of California Berkeley.[4] The group's purpose was to reintroduce deeper philosophical discussions into physics, shifting focus from practical outputs to the deeper inquiries that typified the early days of quantum physics. The perplexing nature of entanglement was a major theme, with its implications of non-local effects, when a change in a state of a subatomic particle is reflected in an entangled particle regardless of the distance separating those particles. This suggested to the group a cosmic oneness suggestive of the ideas being entertained by hippies more generally. This overlap resulted in the Fundamental Fysiks Group members leading workshops at Esalen, explaining how quantum physics could explain the universe in all its grooviness.

Members of the group included Fritjof Capra, who wrote the counterculture classic *The Tao of Physics* in 1975, and Gary Zukav, who wrote the popular science book *The Dancing Wu Li Masters* in 1979, in which he explained quantum physics using metaphors drawn from Buddhism. These books and other talks and works by members of the group are a prominent source of the intertwining of quantum theory with New Age and other pseudoscience ideas that persists to this day. The non-rigorous hand waving is undoubtedly a contributor to our current surge of quantum hype, creating a vague fog of familiarity that makes anything from benign misunderstanding to full-on grift more possible.

It can be challenging to discern the motivation behind some of the hype. Undoubtedly some is simply excitement generated by our enormous leaps forward. Devices are being built today which directly manipulate sub-atomic particles to compute information, like plugging into the fabric of the universe and asking it to solve problems, how can one not be excited by that? Added to that excitement is the truly non-intuitive nature of everything about the sub-atomic world. We are animals who interact with the universe via our senses, but there is nothing to see, smell, touch or hear in the quantum realm.

Tools created by humans spring from our innate desire to be able to accomplish something that is difficult or impossible to attain without that technology existing. They are manifestations of our deepest aspirations and fears, the same source as the stories we tell about the human condition, our mythologies. Any new technology settles into a metaphorical relationship with the mythologies of which it is in some way a recapitulation. The airplane is an expression of the same aspect of humanity that is expressed in the story of Icarus. Scientific theory can cast a wider net, affecting our species' perception of the universe and our place in it. Newtonian physics literally powered the Industrial Revolution, but also made us believe in the watch work universe and the idea of humans as cogs in a machine. Einstein's relativism has metaphorical links to the idea that one's context influences what one perceives.

The cultural metaphors surrounding quantum physics have not yet resolved with much clarity, largely due to the macroscopic limitations of our sensory experience. We assign arbitrary words like spin and color to things that have neither because we don't have better metaphors, but they do not serve to create more clarity in our minds. The idea that an electron is spinning is entirely misleading. Spin would imply the surface of an object revolving around a central axis, but if electrons were spinning, their surface would have to be moving at forty three times the speed of light. That is not what is happening, because electrons have no surface to rotate at all. As Feynman wrote in *The Character of Physical Law,* "Our imagination is stretched to the utmost, not, as in fiction, to imagine things which are not really there, but just to comprehend those things which are there."[5]

This deep confusion is thrust into the center of a cultural and commercial machine that we have constructed since the beginning of the digital era to develop new technologies. This process selects technological innovations with productive potential and creates the marketing messages required to sell investors and customers on its potential, and it has developed into an efficient market mechanism that can take a startup to a massive global corporation. The mix of that well-oiled engine of economic development with a nascent technology based on science that is resistant to our intuitive grasp creates all the ingredients for a hype cycle out of control.

It's safe to say that in 2022 quantum computing was at or near the peak of inflated expectations, but the mistaken beliefs about quantum computing began right at the beginning of the current era of commercialization, or even before it. In Time Magazine's cover story on quantum computing written in 2014[6] as D-Wave was entering the market with its annealer, the confusion is easy to spot. The author quotes from the evaluation of the D-Wave hardware led by Matthias Troyer in which John Martinis and others participated, rightly citing it as a resounding failure for the company. However, in the same article he claims that "Google has been using its D-Wave to, among other things, write software that helps Google Glass tell the difference between when you're blinking and when you're winking."[7]

By 2023, the reporting had dropped any pretense of critical evaluation when Time revisited the topic with another cover story, claiming

> *Quantum's unique ability to crunch stacks of data is already optimizing the routes of thousands of fuel tankers traversing the globe, helping decide which ICU patients require the most urgent care, and mimicking chemical processes at the atomic level to better design new materials. It also promises to supercharge artificial intelligence, with the power to better train algorithms that can finally turn driverless cars and drone taxis into a reality.* [8]

None of this is factually correct. As of 2023 there existed no quantum hardware capable of outperforming a contemporary classical laptop, no practical applications like logistics or AI on quantum computers, and *most definitely* no medical decisions are being made based on quantum software. Worse, the phrase, "quantum's unique ability to crunch...data," perpetuates a misconception that quantum computers will excel at data processing, like Big Data on steroids. This ignores the fact that the machines being built in 2023 have at most a couple hundred bits of working space - quantum bits, yes, but bits nonetheless, and are capable of loading that much data at a time, and no more. Thanks to superposition and entanglement that data can be very high dimensional, meaning it can be very interconnected, but it certainly can't be big. Our supercomputers measure data in the trillions of bits, and aren't going to face any competition from a few hundred qubits anytime soon.

The potent mix of quantum's disembodied nature, pseudoscience wishful thinking, and marketing overreach seems to have actually been eroding the broader public's understanding of quantum computing. At every stage of quantum computing's development there has been this undercurrent of misinformation that has fueled the hype to dangerous levels. Shor's algorithm itself, with its implicit threat to the most powerful encryption methods known at the time, elicited a disproportionate response from the public sector sources of funding attached to our defense industry. Once the immediate perceived threat was dispersed by the slow advance of the hardware, funding receded as well.

The lesson of this dynamic seems to have had the effect of encouraging some individuals from either the science or technology communities to encourage inflated expectations in order to secure their funding. As public sector funding dried up in the 2000s, the idea that quantum computing would have significant commercial benefits began to emerge as the dominant narrative. The Canadian startup D-Wave is in many ways the origin point for the shift to speculative commercial value. Their product launch at the Computer History Museum in Mountain View, California in 2007 is a textbook example of what is sometimes called "science by press conference."

At that demonstration, instead of running something provably hard for classical computers, like factoring, they had their machine solve a sudoku puzzle. An article in Scientific American from 2013, when the backlash against D-Wave was at its peak demonstrates precisely the logical fallacy at the core of quantum hype. The author describes how D-Wave's founder saw his real challenge as the business of quantum computing rather than the science or engineering, and then describes the factoring problem addressed by Shor, explaining that ... "the bottleneck arises because conventional computers store and process information in an either–or fashion, using 'bits' that can each exist in only one of two states, denoted 1 or 0 ... quantum computers, by contrast ... using 'qubits' that can exist as both 1 and 0 at the same time. In principle, they can explore different solutions simultaneously — reducing a multi-year calculation to seconds."[9]

The transition from factoring to the generic "multi-year calculations" is seamless, and inaccurate. There is absolutely no basis to believe that the exponential advantage Shor's algorithm enjoys over classical methods can be generalized in any way, in fact, the search for similar provable advantages has been largely unsuccessful to date. This lack of proven advantage has not slowed the claims, which are greatly aided by the mistaken notion also demonstrated in the quote above, that a quantum computer can "explore different solutions simultaneously."

This misstatement seems to emerge from attempts to explain the practical implications of using superposition and entanglement for computation. In Shor's algorithm, the positive and negative amplitudes of wave functions are used to destructively and constructively interfere with one another, suppressing the wrong answers and amplifying the right ones. This has consistently and persistently been conflated into "trying all possible solutions simultaneously," even by scientists who presumably should know better, like physicist Michio Kaku in a piece on 60 Minutes[10] from 2023. Kaku repeats the "all possible solutions" fallacy in a segment with an animation, suggesting falsely that the quantum computer can trace all possible paths through the maze at once.

With this simultaneity fallacy at its core, quantum computer boosters have been able to make all manner of absurd claims about what quantum computers can do, usually by identifying something hard that classical computers are slowly getting better at, like machine learning, and saying quantum computers will be better at it. Much of this misinformation is meant for consumption by the sources of capital for quantum computing efforts - venture capitalists, the public markets, corporate comptrollers — whatever will keep the lights on in the lab. This dynamic is the source of the most egregious hype, claims that quantum computing will solve the climate crisis[11] or cure cancer[12]. The framing is almost always in the form of "just asking questions" made commonplace by the political consultants, pundits and media in the modern era, couching outrageous fallacies in language like "in the future it may be possible," but the intended effect is the same as in the political sphere: to lead the audience to incorrect conclusions.

The danger posed by misinformation and hype of both the naive, unintentional kind as well as the sort that comes with a pitch for money is that, of course, overinflated expectations inevitably lead to the trough of despair. Bubbles burst, and the backlash against the entire field of quantum computing will lead to some degree of "quantum winter." The phrase is borrowed from the field of artificial intelligence research, which has undergone several cycles of boom and bust, starting with machine translation in the 1960s. Borrowing the idea of a nuclear winter, researchers working on artificial intelligence first coined "AI winter" in the mid-eighties to describe the pattern they had experienced over the preceding years and warned about the possibility it would happen again if expectations exceeded reality.[13] Investment winters unnecessarily slow progress in a field and can destroy careers, and generally should be avoided if at all possible.

We can agree that companies building quantum computers and quantum software should not exaggerate the practical applications of their technologies, but we are left with the question of how to properly understand and explain this stuff. It's not like the experts make it easy for society at large to land on level-headed understanding. In *Fabric of Reality,* David Deutsch wrote that quantum computing "...will be the first technology that allows useful tasks to be performed in collaboration between parallel universes."[14] This quote has surfaced in wild Substack posts speculating that the "Q*" project rumored to have thrown the board and management of OpenAI into conflict is actually a conscious AI running on quantum computing hardware.[15] Even more irresistible New Age bait was inadvertently created by a recent experimental imaging of two entangled photons. The image published in *Nature* needs no further explanation.

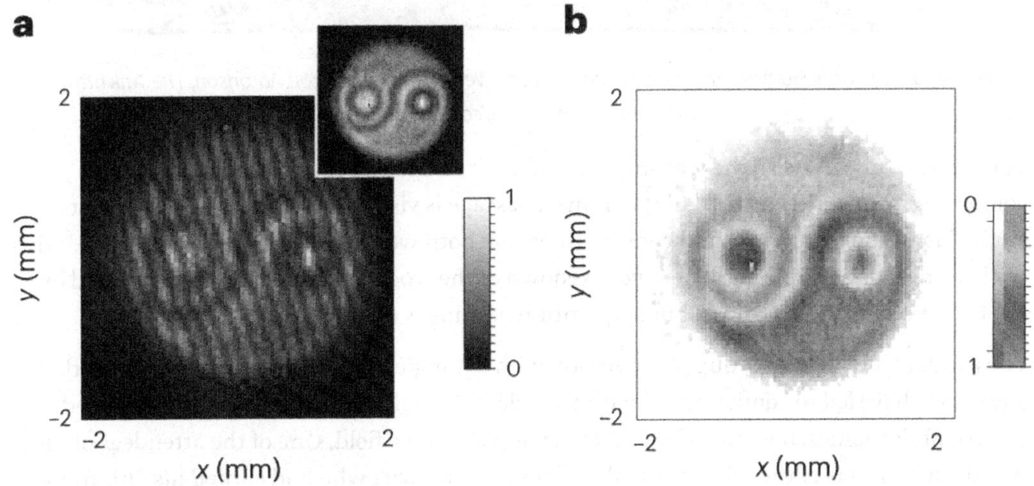

Figure 13.1 Interferometric imaging of entangled photon states. From Zia, D., Dehghan, N., D'Errico, A. et al., "Interferometric imaging of amplitude and phase of spatial biphoton states," Nature Photonics 17, 1009–1016 (2023).

The appeal of the ungraspable core of quantum computing inspires our creative urges in addition to the conspiratorial. Alex Garland's *Devs* series used the common misunderstandings about quantum computing to create a meditation on grief and free will.[16] And of course the "Quantum Realm" featured prominently in the Marvel Cinematic Universe, which abounds with misunderstandings and pure fictional invention, but may also offer some teachable moments about the real science.[17] The need to find those moments when the more knowledgeable can try to take advantage of quantum curiosity to convert some mistruths into understanding has even been given the web comic treatment[18].

Saturday Morning Breakfast Cereal's creator, Zach Weinersmith collaborated with Scott Aaronson on the strip, which hilariously recasts the "birds and bees" talk between a mother and son as her attempt to set him straight about quantum computing.

Figure 13.2 Saturday Morning Breakfast Cereal, by Zach Weinersmith and Scott Aaronson, The Talk https://www.smbc-comics.com/comic/the-talk-3

The comic strip ends with the warning, "if you don't talk to your kids about quantum computing, someone else will." In the frame a passage is visible in a magazine article that reads, "quantum computing and consciousness are both weird, and therefore equivalent." This could be read as a simple and effective takedown of the woo woo proclivities inaugurated by the Fundamental Fysiks Group, but in quantum nothing is ever simply binary.

The Fundamental Fysiks Group's fascination with entanglement, an unproven theory at that stage, may have led to conflation with transcendental meditation and hippy-appropriated eastern philosophies, but it also had practical impact on the field. One of the attendees of the weekly discussions of the group was John Clauser, a physicist who had earned his PhD from the University of California Berkeley. In his studies, Clauser had struggled with quantum physics, having to repeat one advanced class three times before he passed. Working as a postdoc at Lawrence Berkeley National Laboratory, Clauser created an experiment using polarized photons that demonstrated quantum entanglement for the first time. In 2022, Clauser shared the Nobel Prize for his work, and quantum computers performed the same "impossible" operation as part of their boot-up sequence. The hippies didn't just keep entanglement alive as a topic of inquiry during a period when mainstream physics had moved on — they helped create the intellectual environment in which its experimental proof emerged.

And as for consciousness itself — the question that drew the Fysiks Group to entanglement in the first place — the story is more nuanced than "quantum flapdoodle" versus serious science. At the DIWISS Research Institute in Germany, theoretical physicist Joachim Keppler has been pursuing a rigorous scientific framework for understanding consciousness through quantum field theory. Keppler's approach is built not on hand-waving about cosmic oneness but on the mathematical machinery of quantum electrodynamics. His core insight is that the highly synchronized, long-range activity patterns observed in the brain during conscious states display the dynamical characteristics of macroscopic quantum phenomena — specifically, the signatures of self-organized criticality. These patterns, Keppler argues, can be understood as the result of a resonant coupling between the brain and the zero-point field, the omnipresent quantum vacuum that, far from being empty, is a well-characterized ocean of activity described by quantum field theory. It is a serious, experimentally testable hypothesis — and a reminder that the line between the Fysiks Group's intuitions and genuine scientific inquiry is not always as bright as the debunkers would have us believe.

Some amount of humility is required when we are dealing with a subject of such mysterious depth. The history of quantum mechanics is, among other things, a history of ideas that sounded absurd until they were proven correct — from Planck's desperate "hack" to Einstein's spooky action at a distance. Important as it may be to resist our human tendency to characterize anything far beyond our understanding and experience as mere magic, it is also critical not to dismiss lines of inquiry simply because they echo earlier, less rigorous attempts. The quantum realm may not be made of pure wish fulfillment, but it is full of the mysteries of the universe, and it is entirely reasonable to believe that we can both understand those mysteries with greater fidelity, and that our learning will provide enormous benefits in the future.

References

1. Not hyperbole. A startup providing a web interface to Usenet news groups wrote in its S-1 that under its existing business model it was not foreseeable that they would ever produce a profit.
2. `https://www.gartner.com/en/research/methodologies/gartner-hype-cycle`
3. `https://www.scottaaronson.com/democritus/lec14.html`
4. "What Physics Owes the Counterculture," The New York Times, June 19, 2011. `https://www.nytimes.com/2011/06/19/books/review/book-review-how-the-hippies-saved-physics-by-david-kaiser.html`
5. `https://mitpress.mit.edu/9780262560030/the-character-of-physical-law/`

6. "The Quantum Quest for a Revolutionary Computer," Time, February 6, 2014. https://content.time.com/time/subscriber/article/0,33009,2164806,00.html

7. https://content.time.com/time/subscriber/article/0,33009,2164806-6,00.html

8. https://time.com/6249784/quantum-computing-revolution/

9. https://www.scientificamerican.com/article/d-waves-quantum-computer-courts-controversy/

10. https://www.cbsnews.com/news/quantum-computing-google-ibm-advances-60-minutes-transcript/?ftag=CNM-00-10aab7d&linkId=252720335

11. https://eos.org/features/how-quantum-computing-can-tackle-climate-and-energy-challenges

12. https://www.tno.nl/en/digital/digital-innovations/trusted-ict/quantum-computers-cancer-detection/

13. Crevier, Daniel (1993). AI: The Tumultuous Search for Artificial Intelligence. New York, NY: BasicBooks. ISBN 0-465-02997-3.

14. Deutsch, David. The Fabric of Reality. Penguin Books, 1997.

15. https://willdowd.substack.com/p/child-moon-2023#:~:text=According%20to%20the%20Oxford%20physicist,in%20collaboration%20between%20parallel%20universes.%22

16. https://www.nytimes.com/2020/03/04/arts/television/devs-review.html?smid=nytcore-ios-share&referringSource=articleShare

17. https://csferrie.medium.com/is-quantum-bullsh-t-in-marvel-movies-problematic-702a1beafecc

18. https://www.smbc-comics.com/comic/the-talk-3

Subscribe to Deep Engineering

Join thousands of developers and architects who want to understand how software is changing, deepen their expertise, and build systems that last.

Deep Engineering is a weekly expert-led newsletter for experienced practitioners, featuring original analysis, technical interviews, and curated insights on architecture, system design, and modern programming practice.

Scan the QR or visit the link to subscribe for free.

`https://packt.link/deep-engineering-newsletter`

14

The Road to Fault Tolerance

Quantum phenomena do not occur in a Hilbert space. They occur in a laboratory.

— Asher Peres

Architecting reliable systems from unreliable components

Physical manifestations of human ingenuity, no matter how powerful, are never flawless. In operation, the most magnificent machines are subjected to the brutalities of the physical world. Vibrations, heat, cold, strains and stressors affect every screw, seal, gear or axle, and entropy will eventually conquer every one of them, and the machine will fail. Computers, as machines for manipulating symbols, are no different. From the start, one of the critical challenges facing those trying to build a computer was reliability. The teams at the University of Pennsylvania and Princeton's Institute for Advanced Studies both struggled to design their machines around the components available post WWII, particularly the vacuum tubes that implemented the logical gate operations. The tubes were a huge improvement over the purely mechanical parts available to Babbage or the electro-mechanical relays used in previous efforts. They had no moving parts to wear out and they could be switched on and off a thousand times faster. What's more, with no moving parts, their lifespan isn't dictated by the number of times they are switched, unlike anything with a physical mechanism. However, variations in fabrication could lead to filaments in the tubes burning out, or seals breaking and letting air in, or other failures. Some were catastrophic and would cause the machine to halt operations. Worse were failures to operate within the intended parameters but function nonetheless, introducing silent errors into calculations that would result in incorrect results.

Von Neumann, who had established the project to build a computer at Princeton, was a brilliant team-builder, and had recruited some very clever, innovative engineers as part of his team. These engineers, such as Julien Bigelow and Jack Rosenberg, worked to address the issues of operational errors. They tested a thousand of the war surplus 6J6 tubes, selected the worst performing of the lot, then built their system assuming 50% worse performance.[1] Other weak points in the system included unreliable cooling systems and the complications of short- and long-term data storage. Their solutions were not just ingenious, they became the foundation on which the entire information technology industry was built.

No matter how clever or effective, all those engineered solutions were not enough to completely avoid the possibility of a "1" becoming a "0" somewhere in a run comprising millions of individual values, leading to incorrect results. Thankfully, there was a new type of engineering emerging to deal with unavoidable errors, one that could detect and sometimes even correct errors in a bit of information by using other bits of information as a kind of insurance. This work had originated at Bell Labs, where information theory itself was defined and formalized by Claude Shannon in the 1940s. One of his colleagues, Richard Hamming, was an American mathematician who had worked on the Manhattan project before joining Bell Labs in 1946. There, the basics of robust transmission of information over unreliable channels had begun to emerge. The basic techniques included repetition code and parity bits. Repetition code entailed the encoding of the value of a bit in a set number of repeated values. For instance, if your repetition code stipulated five physical bits for each actual bit, a "1" would be encoded as five repeated "1s" in your medium, or "11111." That way, if "11101" was received, you'd know that you could safely ignore the "0" and take the value of the logical bit to be "1." Repetition code only required a majority of the physical bits to retain the original value to successfully deliver the logical bit's value. Parity bits are a more efficient way to detect errors, appending a single bit to the end of a string to make the number of "1s" in the string even (in the case of an even parity, or odd for an odd parity implementation). If a single bit in the string flips, the number of "1s" will change from even to odd, or vice versa, and the value of the parity bit will reveal the error.

Both schemes are capable of detecting when an error occurs, but to Hamming, that was not enough. He wrote, "...surely there is a way of telling where the error is so that the computer can correct the error itself."[2] In 1950, he published a method[3] that used three parity bits along with a four bit 'word,' and a method which could not just detect an error, but correct it. Called Hamming 4,7, or just, "the Hamming code," this was the first error correction code, a technique with broad applications across communications and computation, and an entire field of study unto itself. Over the ensuing seventy years, our techniques for fabricating microprocessors have advanced to the point where there is no need for explicit error correction codes in a modern CPU. However, data storages and communication methods still utilize various descendants of the Hamming codes to ensure their reliability.

With error correction a solved problem in classical information technology, one might presume that the domain of quantum information will be able to rapidly recapitulate prior learning and techniques to deliver error-free operations in short order. That presumption is false, for a number of reasons. Firstly, physical qubits require the most precise engineering ever undertaken. The IAS team was building physical bits with electrical devices as big as a thumb that exploited the behavior of uncountable numbers of electrons in a continuous stream sufficiently powerful to generate heat you could feel on your skin.

In contrast, qubits are created using a single atom. Or a single photon, or electron. In the best case, with superconducting qubits, larger numbers of electrons are behaving as a single quasiparticle, but are so sensitive to their environment they must be kept in a heavily shielded fridge very, very near absolute zero. In all cases, the teams building qubits must do everything they can to separate their devices from the rest of the universe, which wants very much to interact strongly with the particles and quasiparticles, to reincorporate them into the everyday fabric of reality, not isolated for the purposes of performing mathematics. When Feynman and others proposed making computers that used quantum mechanics to carry out logical operations, those who understood what stood in the way of that goal were skeptical, to say the least.

On top of the physical engineering challenges, qubits are vulnerable to two different types of errors. They can suffer from bit flips just like classical bits, where a "0" becomes a "1," or vice versa. Recall, however, that a qubit's state is actually comprised of two values: $a|0\rangle + b|1\rangle$. In addition to a bit flip, it can fall victim to a "phase flip," changing the operator between the amplitudes, so that it becomes $a|0\rangle - b|1\rangle$. Projected onto the Bloch sphere, one can visualize a phase flip as the point on the sphere where the vector "touches" as flipping from "above" the zero axis to below, or vice versa. The amplitudes are preserved, but the computation's results will be invalid just the same.

To compound all of these challenges were the intrinsically weird properties of quantum information that had generated the theorists' excitement in the first place. Recall Stephen Wiesner's quantum currency, impossible to counterfeit. Recall Bennett and Brassard devising a perfectly secure means of quantum communication. Both relied on the quantum mechanical properties of measurement. A two-level quantum system in superposition held a probabilistic combination of both of its states, but when measured that coherent state collapses and the qubit is simply a "1" or a "0." A qubit entangled with another would change its state when the other qubit's state was measured. In short, measurement collapsed the wave function into a deterministic state of "1" or "0." The cat was alive or dead. This poses several challenges for error correction. If one cannot measure a qubit's state, it seems impossible to determine if that state has erroneously been changed. Worse, in order to correct an error one needs to restore the prior information. In quantum information science there is a "no cloning" theorem which finds it impossible to create a copy of a qubit's state. Doing so requires measuring it and measuring it... the cat again.

Quantum error correction was not an active research topic, even after the conference at Endicott, and not even after Deutsch's update to the Church-Turing theorem nor his work with Jozsa on the first quantum computing algorithm. This was a reflection of the prevailing belief that it was an entirely abstract notion unlikely to ever attract the kind of resources needed to even determine if it was feasible. This was captured with astonishing clarity by Asher Peres in his 1985 paper, "Reversible Logic and Quantum Computers,"[4] where he drew the parallel with analog classical computing. This was an approach that theoretically could provide more computational power than digital computers, but there was no efficient error correction scheme known to realize that power in practice.

When Shor's factoring algorithm attracted the attention of the intelligence community, who were *very* interested in seeing if there was a way to make a real code breaking machine, the attitude towards the field abruptly changed. However, there was a general consensus that the kind of complex, precise numerical work required by Shor's algorithm would need error correction, and that the limitations cited above made that very difficult if not impossible to implement. Faced with the skeptical attitude towards quantum error correction, Peter Shor took on the problem himself, and less than a year later, he'd created the first error correcting code for quantum computation.

Shor's workaround to quantum information challenges like the no-cloning theorem was to prepare a circuit that would "smear" a single quantum state out over nine physical qubits, which in aggregate would comprise a single, logical qubit. This logical qubit is a concatenation of a three qubit bit-flip code and a three qubit phase-flip code, making it resistant to either. As you can see from *Figure 14.1* below, a series of quantum gates are applied, entangling three groups of three qubits each to create the logical qubit. Detecting errors after some amount of relevant computational work is achieved by measurement of certain of the qubits from our circuit, which will not only reveal if an arbitrary error was introduced, but also provide the basis for the corrective adjustment to our logical qubit. Those corrections can be carried out with additional gates and finally the resulting qubit state is measured.

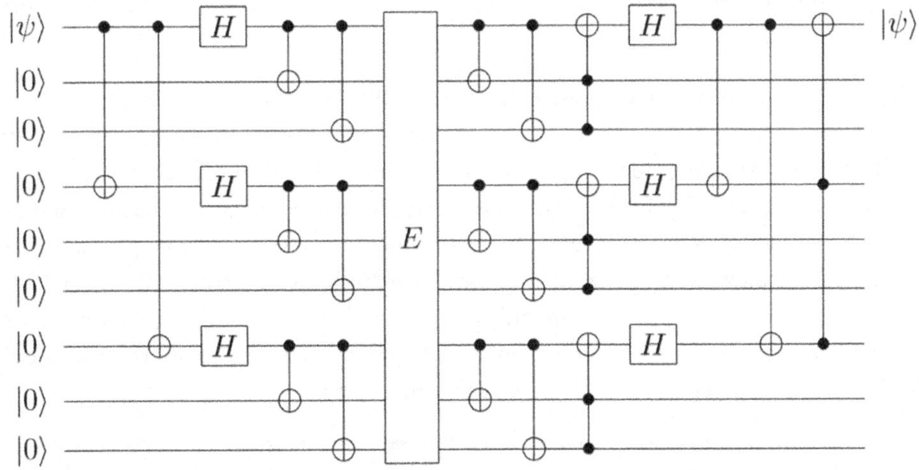

Figure 14.1: Shor's error correction code implemented in a circuit diagram

Shor's algorithm for error correction would work for single qubit errors and was not effective for multiple qubit errors, but it still proved its point. Quantum error correction was feasible despite the unique constraints of quantum information, and that meant that Shor's achievement, the factoring algorithm, might be practically implemented on a real quantum computer. An extraordinary year of accomplishment by Peter Shor, one without parallel in the field of quantum information, and difficult to imagine anyone surpassing in any field.

As is almost always the case with difficult problems, Shor wasn't the only one working on the challenge of error correction. Certain personalities seem drawn to the most vexing of puzzles, and the field of quantum computing has an abundance of just those types of people. Andrew Steane, a research fellow at Oxford University in 1995, had submitted a paper to Physical Review Letters in October, only to find out about Shor's work during its evaluation. Steane's error correction approach was similar to Shor's, adapting aspects of Hamming's classical error correction to the constraints of quantum information to create a single logical qubit out of seven physical qubits. The slight increase in efficiency arose from Steane's use of a single ancillary qubit for syndrome detection versus Shor's multiple error detection qubits, but both could detect and correct a single qubit bit- or phase-flip error.

By the time Physical Review Letters published Steane's paper, Shor had found a collaborator, Robert Calderbank, a fellow researcher at Bell Labs, and they had submitted a paper to Physical Review A, presenting a generalized theory of quantum error correction based on Shor's technique, titled "Good quantum error-correcting codes exist."[5] The research community, recognizing the similarities between the two parallel efforts, came to refer to the approach collectively as Calderbank-Shor-Steane codes, or CSS for short. By 1998, the first experimental proof of quantum error correction was implemented on three NMR qubits by a group working at MIT. [6]

Following these leaps forward, however, quantum error correction stalled along with the field's ability to create larger numbers of higher quality qubits. Progress slowed, but, like the hardware efforts, it did not stop. Additional work to generalize the approaches into use resulted in formalizing "additive" codes, where multiple techniques were concatenated to create a holistic error correction system. An improvement on Shor's original approach was published by Dave Bacon in Physical Review Letters in 2006, using a two-dimensional arrays to further reduce the overhead of the encoding. Additional experimental implementations of CSS codes, and derivations of CSS such as color codes, have been carried out in various qubit modalities such as trapped ions and superconducting qubits.

Shortcomings remained in these types of codes derived from classical methods of error correction, however, namely vulnerability to complex multi-qubit errors or combined errors on a single qubit. CSS codes are also only effective with Clifford gates and not the Toffoli gate, which is critical for Shor's and Grover's algorithms and many quantum simulations. An alternative school of thought began to emerge in 1997, when Alexei Kitaev, a brilliant physicist then at the Landau Institute for Theoretical Physics in Russia, proposed a method for projecting qubits states onto a lattice, seen in *Figure 14.2* below, whose edges wrap around to join one another, forming a torus.

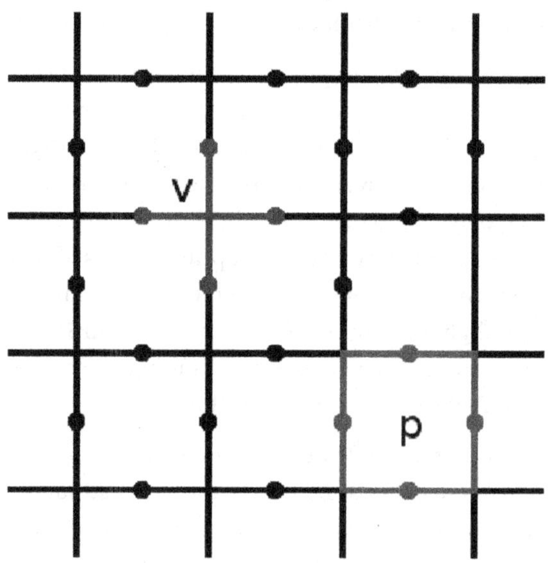

Figure 14.2: By Woottonjames - Own work, CC BY 3.0, `https://commons.wikimedia.org/w/index.php?curid=11823316`

Each intersection on the lattice is a vertex, one of which is labeled v in *Figure 14.2*, and each square in the lattice is known as a plaquette, labeled p. The vertices and plaquettes act as stabilizers for the qubit states, depicted as the dots on each side of a plaquette. When an error occurs in any qubit, the stabilizer's state will reflect that error, creating a quasiparticle called an anyon. The resulting anyons will travel around the torus, seen in *Figure 14.3*, erasing each other, thereby undoing the original error. The toric code can be used to encode two logical qubits from twenty-one physical qubits.

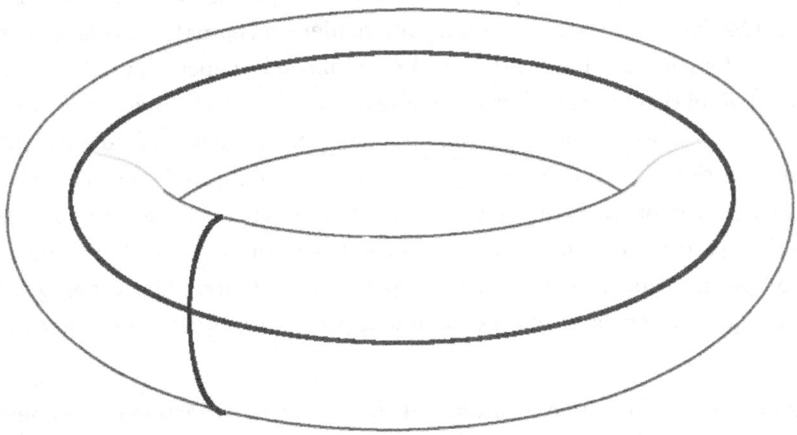

Figure 14.3: By Woottonjames - Own work, CC BY 3.0, `https://commons.wikimedia.org/w/index.php?curid=11824034`

The toric code was the first topological error correction scheme, an approach that, despite its enormous complexity, began to gather more attention and effort due to its highly resilient nature. Mathematical topologies are ways to describe connectedness of complex surfaces, and the relative distances that exist among elements described within that surface, or lattice. The nature of topologies is such that those relationships may not change even if numerical values of the elements do. An analogy that is sometimes used is that a knot in a string is a topology - rotate the knot, move it in space, and the relationship of the string looped around itself does not change. Kitaev had seen the potential for powerful information preserving functionality, something like the use of knots for recording information by the Incan culture as long as 4500 years ago. The oldest surviving quipu dates to the first millennium C.E., and that longevity is imparted by the structure of the knots, their intricate structures encoding information that holds back the erasure of nature and the elements.

In fact, Kitaev's early work connecting topologies to quantum error correction is the genesis for the idea of topological qubits of the kind that Microsoft tried to create starting in 2012. The idea of a qubits with topological protection built into the material out of which it is fabricated was powerfully attractive, but such a material would require harnessing an exotic particle called a Majorana fermion, whose existence remains, to this day, merely theoretical. Topological error correction schemes can be implemented, however, on run-of-the-mill qubits in existence today. A number of small-scale experiments have instantiated aspects of the toric code and topological techniques that it spawned. After the toric code in 1997, Kitaev went on the develop surface code, a more robust and generalized error correction scheme that in many ways remains the gold standard to this day.

As the techniques of additive and topological codes developed, Israeli researchers Dorit Aharanov and Michael Ben-Or were working on broader conceptual proofs for the efficacy of error correction for computation in general. They published a paper in 2008 which established the theoretical proof that all this effort was going to be worth the trouble. Their threshold theorem held that, if the detection and correction of errors could be carried out faster than the errors inherent in the system could propagate, the lifetime of a qubit's state could be extended to arbitrary lengths. In other words, an error correction scheme efficient enough to stay ahead of the underlying error rate of the system on which it ran would provide fault-tolerant quantum computing, capable of long-running, precise calculations. Like running Shor's algorithm to analyze a very large number to find its prime factors, and breaking a code such as RSA.

The combination of the threshold theorem and effective error correction techniques would seem to have created a clear path to quantum computing applications that can solve some very difficult problems efficiently, thereby creating economic value. However, the map, in this case, is not the territory. The performance of the surface code suggests that an error rate as high as 1% might still fulfill the threshold theorem to produce a fault-tolerant quantum computer, but achieving the large numbers of qubits required that also operate at the necessary fidelities has proven to be immensely difficult. Systems with larger numbers of qubits, like IBM's 1121 qubit "Condor" chip, have struggled with high levels of noise. Worse, IBM's strategy to reduce noise produced a chip layout they call "heavy hex," which limits connectivity among qubits to two or three neighbors. This means that large, complex circuits with a lot of entanglement require huge numbers of SWAP gates to "move" a particular state to a where it can be entangled with the target state, and all these additional gates rapidly run out the coherence time of the system. Implementing the overhead of the surface code in such an environment is impossible.

Grappling with the monumental engineering challenges of quantum computing devices has inspired many creative attempts to find some middle ground between the abstract techniques of quantum error correction and the messy reality of physical qubit implementations. One such innovative idea also bears the name Kitaev, though it is not another iteration of his work on topologies. Instead, Kitaev, working for Microsoft at the time, worked with a fellow Microsoft researcher, Daniel Gottesman and John Preskill from Caltech, to enlist "unused" energy states of qubits to carry out error correction.

In a qubit, two energy states are mapped to the |0> and |1> basis somewhat arbitrarily, based on the relative ease with which those states can be initialized, and reliably measured. Aside from those discrete states, the quantum system of a qubit has an infinite Hilbert space that is "unused." GKP error correction (after Gottesman, Kitaev, and Preskill) the bosonic states within this space are used for continuous variables that act in the place of stabilizers or ancillary qubits in other schemes. GKP, like the toric code, has inspired variations on the approach that, collectively, are referred to as bosonic codes.

Despite the promise of bosonic and surface code approaches to error correction, experimental implementations were slower coming than the initial demonstration of additive codes in the style of CSS. For surface codes the number of physical qubits and the fidelity and number of connections among those qubits proved quite demanding for hardware being built in the 2010s. Bosonic codes rely on hardware design for a successful implementation, something that requires tight collaboration between theorists and experimentalists to realize. That type of collaboration typifies the groups working at Yale under Girvin, Schoelkopf and Devoret, the same team that spawned the Yale gang of Gambetta, Rigetti, Chow and others. In 2016, the Yale group published a successful demonstration of a bosonic code on a superconducting device.[7] The qubit was redundantly encoded in higher dimensions, using the quantum states at higher levels than the 0 and 1 states, so that if a bit flip error occurred those upper levels would serve as a space for error detection. The experiment goes further than merely detection, implementing the full cycle of encoding a state, monitoring it, decoding and, if needed, correcting it. Most exciting, the team demonstrated "break-even," or, in terms of the threshold theorem, an ability to correct errors faster than they propagate in the system, extending the effective lifespan of the qubit's state. The paper reported a total lifetime of 320 microseconds, 20 times longer than the coherence time of the physical transmon.

The bosonic encoding is of a Schrödinger superposition of the qubit's state, and in honor of his *gedankenexperiment*, this type of implementation has been dubbed "cat qubits." The work has continued at Yale, with a commercial spin out, Quantum Circuits, Inc., launched in 2015, set to launch its first public systems in 2025. Cat qubits have attracted the attention of several other groups, including Nord Quantique, spun out of the University of Sherbrooke, where another Yale alum, Alexandre Blais, has been building an impressive group. In France, Alice & Bob are also working on cat qubits, with very impressive results such as protection from bit flips measured in minutes, not microseconds, presented at the American Physical Society meeting in March 2024. Their public roadmap projects 100 logical qubits operational by 2030. Amazon's quantum R&D team, co-located on the Caltech campus and led by the experimentalist Oskar Painter collaborating with the theorist Fernando Brandao is reportedly also pursuing cat qubits, with no public roadmap or accessible systems at this time.

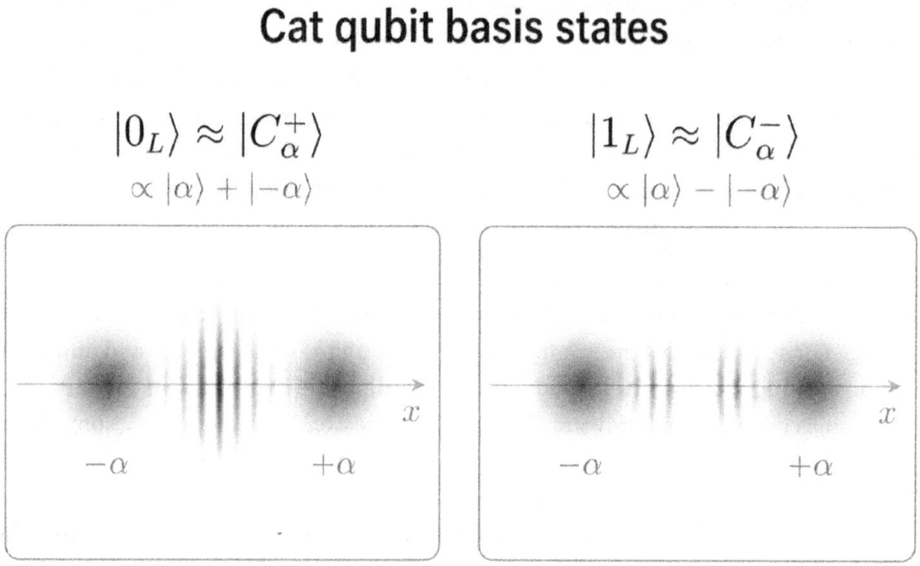

Figure 14.4: The basis states of a cat qubit, created by the author using ChatGPT 5.5

Surface codes are unforgiving on hardware in part because they, unlike cat qubits, aren't dependent on specific features in the device. A sufficient number of physical qubits with sufficient connectivity (a lattice topology is ideal) and sufficient fidelity is all it takes. The word "sufficient" is doing a lot of work here. Trapped ion qubits have excellent fidelity and all-to-all connectivity, however, their operations are so slow that the effective "break-even" point is higher, giving the edge to superconducting systems with larger qubit counts. Unfortunately, large numbers of superconducting qubits tend to have lower fidelity on both one- and two-qubit gates, and often trade off connectivity to get slightly better fidelities, such as IBM's "heavy hex" layout mentioned above. In spite of all this, the team at Google, using their 72 qubit system, a successor to the "Supremacy" device from 2019, demonstrated the first successful surface code error correction in 2022[8], implementing a distance-5 code using 49 physical qubits, seen in *Figure 14.5*.

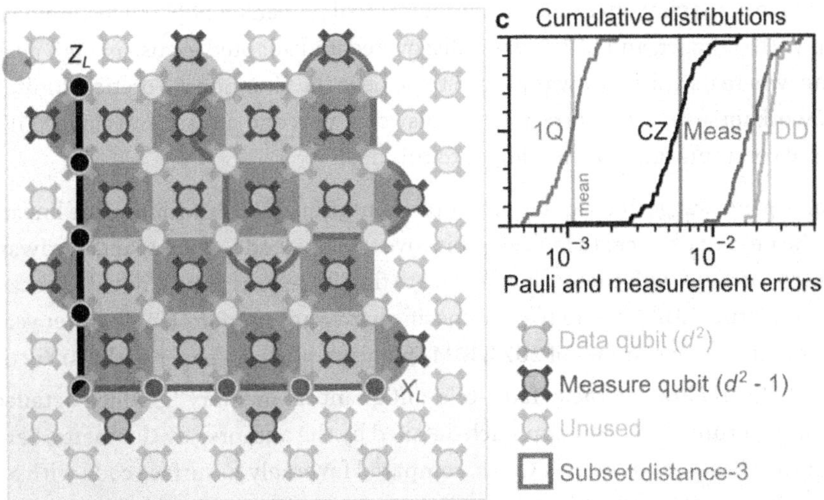

Figure 14.5: Google's distance-5 surface code implementation `https://arxiv.org/pdf/2207.06431#page8`

Google's experiment ran the code through 25 cycles of encoding, detection and correction, and found that the distance-5 code outperformed a distance-3 code implemented on 17 qubits. This was the first experimental proof that lower error rates could be obtained with higher distance surface codes on larger numbers of qubits. Undoubtedly a major validation of the technique, but by no means did it mean error correction was a solved problem.

The distance of a surface code implementation reflects the degree of redundancy in the encoding, corresponding to the number of errors that can be detected and corrected without measuring the data qubits. In order to maintain a logical qubit, you'd need to re-encode the qubit states before exceeding that number of errors. Measuring and re-encoding will impose a classical overhead on your quantum computation, threatening any potential quantum speed up. Presuming physical qubits remain in the realm of frequent errors, a large distance will be needed for complex tasks like Shor's or quantum simulations. A very large distance will require a very, very large ratio of physical qubits to logical qubits, and all indications are that we are quite a ways from 10s or 100s of thousands of high quality qubits.

Faced with this scaling problem, theorists working on quantum error correction looked again to classical methods for possible solutions. In 2013, Daniel Gottesman published a paper[9] that suggested yet another approach, adapting Low-Density Parity Checking, or LDPC, to a quantum information regime. LDPC was discovered by Robert G. Gallager as part of his doctoral thesis[10] at MIT, in 1960. Using sparse measurement to detect errors, Gallager's code promised highly efficient encoding and detection, but did not generate significant interest at the time due to constraints in available computational power. By 1995 the technology had caught up with Gallager, and LDPC was rediscovered and adapted for use in high performance computing, where its efficiency was proven to be near the Shannon theoretical limit. LDPC became a workhorse for error correction in noisy channels and, in fact, is used in WiFi and other digital communication technologies to this day.

Gottesman's paper suggested a quantum implementation of LDPC would scale linearly, greatly reducing the ratio of physical to logical qubits over surface code. Once again, hardware took some time to catch up with theory, in part due to fidelity requirements, but also because, like cat qubits, quantum LDPC would require specific hardware support. Finally, after years of building excitement, in summer of 2023, IBM Quantum posted a paper to the ArXiv titled, "High threshold and low-overhead fault-tolerant quantum memory[11]," which detailed their strategy for quantum LDPC. The approach detailed by the authors, Jay Gambetta, Sergei Bravyi, Dmitri Maslov, and Andrew Cross, compared favorably to surface code with a threshold close to 1%, distance-12, with 12 logical qubits composed of 288 physical qubits maintained for 1 million cycles of syndrome detection and correction. The LDPC codes would only protect Clifford gates, which can be efficiently simulated by classical means, however, IBM described in another 2023 paper[12] clever workarounds that rely on the precision of their custom-built control hardware providing virtually error-free z-rotations, providing a universal set of gates.

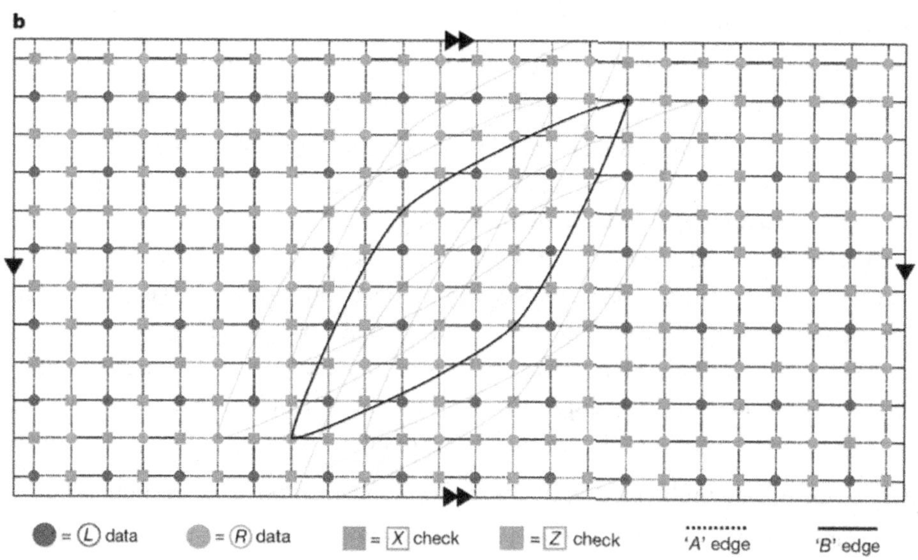

Figure 14.6: IBM's LDPC error code, also known as Bivariate Bicycle, or "gross" code

Amazing result, with just one problem (there always is one, isn't there?) — the 288 physical qubits in the paper are laid out in a bi-planar chip architecture that is completely unlike the devices IBM has been making. In practical terms this could mean a delay of two to three years or more to allow the chip design, fabrication, characterization and packaging teams to retool everything they've been doing to this new architectural approach. Other qubit modalities, those not requiring ingenious and precise chip design and fab, might actually deliver IBM's LDPC scheme at scale before they do.

In December of 2023, the groups working under Mihail Lukin at Harvard and Vladan Vuletic at MIT, published a paper that represented the first step to do exactly that. Lukin's and Vuletic's used their neutral atom array of 256 atoms to realize 48 logical qubits following the broad strokes of IBM's LDPC technique, right down to using the equivalent of z-rotations, called transversal operations in this context, to deliver a universal set of gates. Controlling the atoms in the vacuum chamber, they were able to build a kind of Von Neumann machine floating in their lattice of lasers. As seen in *Figure 14.6*, the scheme involves creating several special-purpose zones for entangling qubits, performing two-qubit gates, storing logical qubits, ancilla qubits, syndrome detection, and qubit readout.

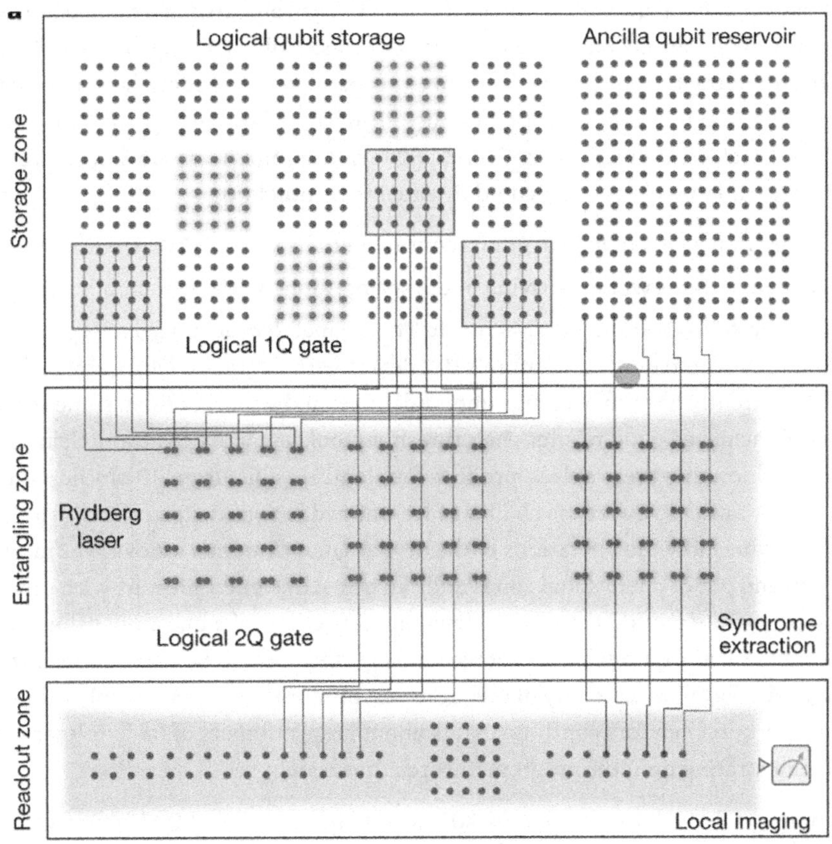

Figure 14.7: virtual architecture of neutral atoms used to realize logical qubits, Lukin, Vuletic et al https://www.nature.com/articles/s41586-023-06927-3

It must be acknowledged that the experiment only performed the phases of fault tolerant operation in isolation, that is, they encoded logical qubits, detected errors, and performed protected gate operations, but they did not link those steps into an end-to-end process. This will undoubtedly be the next priority for Vuletic and Lukin, along with the porting of the technique to the products of QuEra, a commercial neutral atom quantum computer company, which they co-founded.

Despite the partial result of the paper, its publication in Nature caused somewhat more of a stir than IBM's papers early that year. Partly this was due to the fact that IBM was referring to future hardware, forecasting functionality and performance rather than demonstrating it as the neutral atom paper had. More importantly, though, was the sheer scale of the work done by MIT and Harvard. It is generally accepted that classical simulations of quantum computers cannot scale beyond 50 qubits when running demanding algorithms with a lot of entanglement, like Shor's. That the neutral atom experiment was able to produce 48 logical qubits out of 256 physical qubits implies that QuEra may be able to deliver a machine that truly can perform operations impossible to carry out classically. IBM's 12 logical qubits' calculations, in contrast. would be well within the capability of a modern laptop to simulate. Recognizing this, IBM is also talking about a custom-designed fridge that will house five of these devices in a cluster, effectively delivering 60 logical qubits. This, too, is a future deliverable with unknown risks that could threaten their timeline.

Progress towards a new era of logical qubits continued to be made in 2024, with a peer reviewed version of IBM's work[13] arriving in the spring, along with a pre-publication from Quantinuum[14] in collaboration with Microsoft. Their paper demonstrated a mixture of techniques implemented on Quantinuum's H-2 device with 32 trapped ion qubits. Quantinuum currently has the best two-qubit fidelities on the market, so in some sense it's not surprising they achieved good results, their threshold took less work to attain. Using a variety of Steane-like codes, they were able to produce four logical qubits from 30 physical qubits, an impressively low ratio. However the challenge for trapped ions remains a solution for scaling that doesn't further slow the operations of the device. Quantinuum's innovative "racetrack" moves ions in and out of operational zones and can be used to entangle ions with others in different traps, getting around the thermodynamic restrictions of a one-dimensional trap, but at speeds that make any quantum advantage unlikely, and seem to be stuck in the dozens of qubits range. Finally, in the category of big promises, photonic vendors, notably Psiquantum, and several spin qubit vendors continue to talk about large numbers of fault-tolerant qubits, without demonstrating tangible progress towards those goals.

Perhaps a naive read of the market in mid-2024 would place IBM and QuEra firmly in the lead in terms of tangible results and roadmaps leading to more fault-tolerant qubits than can be simulated. Important to note, however, that only surface code, today, can deliver protected

Toffoli gate operations, through a very convoluted set of operations called "magic state distillation." The LDPC codes from IBM and QuEra provide protection for Cliffords, with rotations rounding out the universal set of operations, instead of Toffoli gates. This is subtly moving the goal post on what constitutes a logical qubit. It may prove sufficient to realize some level of commercial value but at this point it's yet another known unknown to contend with, refusing to resolve the future of the technology with any certainty. What does seem certain is that the predictions that NISQ devices would be unable to produce commercial value were on the mark. Prominent leaders of software companies once bullish on variational techniques have expressed increasing skepticism, with the CEO of Qunasys, Tennin Yan, saying on stage at Q2B Paris in 2023 that "VQE is dead."[15]

And then, in December 2024, Google unveiled Willow.

The 105-qubit superconducting chip, whose technical results had been quietly posted to the ArXiv that August,[16] was formally published in *Nature* alongside a publicity campaign that recalled the fanfare of the 2019 supremacy announcement.[17] But where the supremacy claim had been met with legitimate skepticism, Willow's central result was far harder to argue with. The team had implemented surface codes at distances 3, 5, and 7, and for the first time, each step up in code distance cut the logical error rate roughly in half — an error suppression factor of 2.14, well below the threshold. The distance-7 logical qubit, composed of 101 physical qubits, achieved an error rate of just 0.143% per correction cycle. Most strikingly, the encoded logical qubit outlived its best constituent physical qubit by a factor of 2.4. The error correction overhead was not just paying for itself — it was delivering a net gain. This was the milestone the field had been chasing since Shor's first code in 1995: experimental proof that adding more qubits to the error correction scheme made things better, not worse.

Google also ran a random circuit sampling benchmark on Willow, estimating that their chip completed in under five minutes a computation that would take the Frontier supercomputer approximately 10^25 years — a margin so vast it exceeded the age of the universe by fifteen orders of magnitude. Critics were quick to note, as they had in 2019, that random circuit sampling has no practical application. But the error correction result needed no such caveats. It was a clean, reproducible demonstration that surface code worked as the theorists had promised.

Almost exactly a year later, Quantinuum answered with a machine that redefined the competition. Their Helios processor, announced in November 2025, carried 98 physical qubits — but they were barium ions, not the ytterbium that had powered all of Quantinuum's prior systems.[18] Barium's optical transitions sit at accessible visible wavelengths, enabling gate fidelities that set new records: 99.9975% on single-qubit gates and 99.921% on two-qubit gates, averaged across all operational zones. The racetrack architecture that had constrained the H-2 was gone, replaced by a fundamentally new design built around the world's first commercial ion junction — a four-way intersection connecting quantum logic zones to a ring-shaped storage loop, creating a separation between memory, cache, and processing zones that bore a striking resemblance to classical computer architecture.

The error correction results, published in early 2026, were even more remarkable than the hardware.[19] Using high-rate "Iceberg" codes uniquely suited to trapped ions' all-to-all connectivity, the team produced 48 fully error-corrected logical qubits from 98 physical qubits — a ratio of just 2:1, a figure that would have been dismissed as fantasy only a few years earlier. Surface code implementations on superconducting chips typically require ratios closer to 1,000:1 for equivalent error suppression. Every logical qubit outperformed its physical counterpart, some by factors of 100. Using error-detecting codes, the team went further still, entangling 94 logical qubits in a single GHZ state at nearly 95% fidelity. These were not the twelve logical qubits they had committed to publicly — this was something qualitatively different, a result that forced the entire field to reconsider the scaling assumptions that had favored superconducting architectures.

Taken together, Willow and Helios represented the most significant year for quantum error correction since Shor and Steane's papers three decades earlier. Two radically different qubit modalities had independently demonstrated that fault tolerance was not merely theoretically possible but experimentally achievable, each with its own advantages and limitations. Surface code on superconducting qubits offered speed and a clear path through established semiconductor fabrication. Trapped ions offered precision and encoding efficiency that seemed almost unfair. Neither had yet produced a machine capable of solving a problem that mattered to anyone outside the field. But the road to fault tolerance, which had seemed interminable for so long, now had landmarks on it — and the distance between them was closing.

As impressive as the results from both companies, they also pointed out an emerging debate and semantic dance around the definition of what is meant by "logical" qubits, precisely. Intrinsic to the field of error correction is the notion of "distance," which is a measure of robustness. Formally, a distance-d code can detect up to d-1 errors and correct up to $(d\text{-}1)/2$. A way to think about it is as witnesses to a crime – if one of three independent testimonies gives a different account than the other two, you can use the majority to construct an accurate picture of the crime. The more witnesses, the more erroneous accounts your reconstruction of the events can withstand. Google's surface code provided d=7 on Willow, while the Helios experiment traded robustness for efficiency. The 2:1 ratio of physical qubits to logical qubits sounds great, but d=4 means errors on that platform would have more destructive potential to the calculation.

More nuance is always hidden in the details of these fault tolerant results – some do post-selection, harvesting noisy results and cleaning them up after the fact, which in terms of practical applications is frankly a bit of a cheat. More subtle is the degree to which a particular demonstration covers the entire logical qubit life cycle from detecting errors to correcting them while maintaining the intended algorithmic process. The impressive 94 error-detected qubit entangled GHZ state on Helios, for example, was using d=2 codes to only detect errors, not correct them. Impressive, of course, but it doesn't pass the production-level operational test.

In the coming years error correction results and logical qubit demonstrations and claims will swell from a trickle to a torrent, and decoding the truth within the chest-puffing marketing posturing will be an increasingly challenging task.

References

1. Bigelow, Pomerene, Slutz, and Ware, "Interim Progress Report," pp. 82-83.
2. https://amturing.acm.org/info/hamming_1000652.cfm
3. R. W. Hamming, "Error detecting and error correcting codes," in The Bell System Technical Journal, vol. 29, no. 2, pp. 147-160, April 1950, doi: 10.1002/j.1538-7305.1950.tb00463.x.
4. Peres, Asher. "Reversible Logic and Quantum Computers." Phys. Rev. A 32 (December 1985): 3266–76. https://doi.org/10.1103/PhysRevA.32.3266
5. Calderbank, A. R. and Shor, Peter W. "Good quantum error-correcting codes exist." Phys. Rev. A 54, no. 2 (August 1996): 1098–1105. https://doi.org/10.1103/PhysRevA.54.1098
6. https://arxiv.org/pdf/quant-ph/9802018
7. Ofek, Nissim; Petrenko, Andrei; Heeres, Reinier; Reinhold, Philip; Leghtas, Zaki; Vlastakis, Brian; Liu, Yehan; Frunzio, Luigi; Girvin, S. M.; Jiang, L.; Mirrahimi, Mazyar. "Extending the lifetime of a quantum bit with error correction in superconducting circuits." Nature 536, no. 7617 (August 2016): 441–445.
8. https://arxiv.org/pdf/2207.06431
9. https://arxiv.org/abs/1310.2984
10. https://dspace.mit.edu/handle/1721.1/11804
11. https://arxiv.org/abs/2308.07915
12. https://arxiv.org/abs/2310.16982
13. Bravyi, S., Cross, A.W., Gambetta, J.M. et al. "High-threshold and low-overhead fault-tolerant quantum memory." Nature 627 (2024): 778–782. https://doi.org/10.1038/s41586-024-07107-7

14. https://arxiv.org/abs/2404.02280
15. https://q2b.qcware.com/session/q2b23-paris-beyond-vqe-advancing-quantum-computing-applicability/
16. https://arxiv.org/html/2408.13687v1#abstract
17. Acharya, R. et al. (Google Quantum AI). "Quantum error correction below the surface code threshold." Nature 638 (2025): 920–926. https://doi.org/10.1038/s41586-024-08449-y
18. Delaney, R.D. et al. "Helios: A 98-qubit trapped-ion quantum computer." arXiv: 2511.05465 (November 2025). https://arxiv.org/abs/2511.05465
19. Quantinuum. "Computing with many encoded logical qubits beyond break-even." arXiv:2602.22211 (February 2026). https://arxiv.org/abs/2602.22211

15

The Quantum Horizon

> *We tend to overestimate the effect of a technology in the short run and underestimate the effect in the long run.*
>
> — *Roy Amara*

Conjuring up a new reality for us all

2025 was, by any measure, a banner year for error correction, and the many high-profile breakthrough stories also broke through into the general public awareness, specifically, the public who trade stocks. IonQ was the first "pure play" quantum technology company to be publicly listed, arriving on the NASDAQ market in 2021 via a SPAC, and by 2025 there were four quantum stocks in total: IonQ, Rigetti, D-Wave and Quantum Computing, Inc. Despite the small and somewhat arbitrary assortment of choices, the markets responded to indications of real advancements vigorously, investing $1.2B in the first three quarters of 2025 alone, more than twice the total number for all of 2024. Individual stocks soared, with Rigetti and D-Wave seeing peak gains of over 6000% and 3000% before settling down to a "modest" +233% and +458% respectively. Astonishingly, it is estimated that Google's surface code results on the Willow chip, announced in December of 2024, moved more than $100B in market capitalization across quantum-exposed ticker symbols.

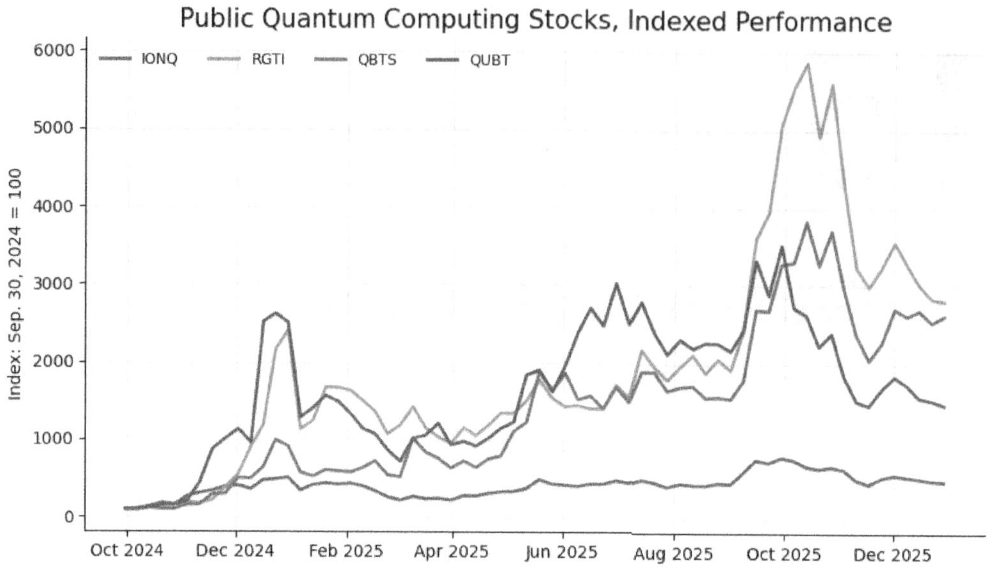

Figure 15.1: quantum technology stock price performance indexed to September 30, 2024, source: Yahoo Finance

Markets aside, 2025 saw notable advances in quantum information technology along many other dimensions. As already mentioned, the Nobel Prize in physics was awarded to John Clarke, Michel Devoret, and John Martinis for their work, starting in the 1980s, demonstrating macroscopic quantum tunneling behavior in superconducting circuits. In other words, for laying the foundation for superconducting qubits. Clarke had been, according to Brian Josephson, convinced from the very start of his junctions' potential value. Devoret formed one of the three legs of the mighty Yale Quantum Institute along with Girvin and Schoelkopf and helped create the transmon architecture, and Martinis, of course, had been the experimentalist who built the Sycamore-powered system behind the Google Supremacy experiment. The Nobel was read as a full-throated endorsement of the accomplishments of the field as a whole, a validation by the academy that Feynman had been proven correct, that this was a good problem to solve.

On top of that, 2025 was also designated the United Nations' International Year of Quantum Science and Technology, as a celebration of quantum mechanics' 100th anniversary. In June of 1925, Werner Heisenberg's hay fever drove him to the clean air of the island of Helgoland in the North Sea, where his struggles to resolve various theoretical physics challenges led to the formulation of the Uncertainty Principle. In June of 2025, Yale University and the Max Planck Institute held a conference on Helgoland, a celebration of the "birth" of quantum mechanics. It featured an expansive agenda that ranged from foundational interpretation of quantum mechanics to the emergence of quantum computing, communication, and metrology technologies, and featured a handful of Nobel Laureates. The event was auspicious enough to inspire an effort to capture it in a documentary film, currently in post-production (full disclosure, I am the filmmaker). The Helgoland documentary was not the only effort to capture and communicate the trajectory of quantum technology; the Quantum Insider, a specialty analyst firm, announced they were making Our Quantum Future, a group affiliated with the World Economic Forum announced another, and word got out that IBM was creating its own documentary to provide a narrative of its efforts.

The narrative arc beginning with Google's supremacy experiment on Sycamore seemed to be building to a climax. All signs pointed to utter triumph and the arrival, finally and completely, of quantum technology on the world stage. Or seemed to. The reality was quite a bit more complex, like everything quantum-related inevitably is.

As I write these words in March, 2026, there are no quantum applications. There are no production workloads running on quantum hardware. There is no economic value in quantum devices. And any claims to the contrary are either delusional or an attempt to create a marketing narrative that doesn't reflect the actual reality.

That's the brutal truth of quantum information technology. Accurately parsing press releases and analyzing the truth behind research breakthroughs remains dependent on a deep understanding of the intricacies of the field itself, a familiarity with the fundamental building blocks of quantum information science and technology at a level that hasn't been required in classical information technology for many decades.

Promise remains. If anything, there's more promise and potential than ever before — recognition of the enormous power of this wave that is building. But that wave is still building. It hasn't crested; it's not crashing on the shore. It's not having societal impact at scale the way we expect any new digital technological innovation to do. The gap between quantum's promise and its present comes down to physics, economics, and time — and none of those are moving at the same pace.

The reasons for this are myriad and vary tremendously from one effort to another, even one qubit modality to another, even one variation on a qubit modality to another variation. But in general, they boil down to a combination of engineering challenges and science challenges.

Engineering challenges are often related to purity and precision — purity of materials that are going into the devices that are being made, precision of things like the placement of a single atom vacancy in a diamond matrix, which, if it's off by one atom, might cause less than optimal behavior in the device that's fabricated from that material. There are many such engineering challenges, but the thing that unifies them, characterizes them, is the mind-bogglingly tight tolerances. There's an anecdote about an iPhone model where Apple and their manufacturers got the tolerances down between the parts so low that it was effectively waterproof just by virtue of how tightly the pieces connected. Those are micrometer tolerances. When TSMC and ASML collaborate on a chip design and fabrication, those are nanometer tolerances. These are atomic tolerances. In a kind of inversion of Ray and Charles Eames Powers of Ten, the scale of a single atom is yet another order of magnitude smaller.

On the scientific front, despite everyone's best hopes, there are real scientific challenges that underpin many of the objectives that are being characterized as simply engineering. Scale, for example. One approach to scale is via transduction — the process of taking a qubit state from one quantum processor and transporting it, communicating it, to another quantum processor. For example, a superconducting qubit state, which is interacted with at microwave frequencies in the bottom of the dilution fridge, being converted by a process of transduction into optical frequencies, turned into what is often referred to as a flying qubit, sent across a room to another dilution refrigerator where it becomes a qubit in that fridge's QPU. "Flying qubits" are simply quantum information encoded in photons and sent down a fiber optic wire.

Given there are quantum computers that use photonic qubits such as PsiQuantum, Xanadu, Quandela or ORCA, this might seem like a solved problem. The trick here lies in the fact that microwave frequencies differ from optical frequencies by almost five orders of magnitude, and trying to step up that far or step down that far produces massive challenges in plucking the signal from the noise. Transduction technologies do exist, but they still tend to be fairly lossy, with the best results recently crossing the 80% mark — above the critical 50% no-cloning threshold, but still far from the near-unity fidelity that production-scale quantum networking would require.

And many other aspects, once we dig into them, reveal the same kind of challenge — not just an engineering problem, but actually some as-yet-unknown scientific discovery that's going to potentially be required for a solution.

A good example is photonic entanglement. The traditional method takes two single photon emitters and pumps out two photons at a time, with their configuration and operation designed to create entanglement, which is required for a two-qubit gate where photons are your qubits. However, photons are bosonic and are highly resistant to interacting with one another. That's why we don't have the same requirement for extreme millikelvin temperatures to operate photonic qubits, which can be relatively hot compared to superconducting qubits, because of that resistance to interaction with their environment. So the entanglement is generated probabilistically, and you have to pump out large numbers of photon pairs to get some entangled pairs on which you can then operate. PsiQuantum addresses this by having entanglement factories that are constantly pumping out photon pairs and storing them for when they're going to be used. That creates an enormous operational overhead — the system that PsiQuantum imagines building for a million photonic qubits would require an entire data center of equipment to be generating and storing these entangled pairs for operations.

Newer approaches are being explored which combine spin qubits of silicon with a single photon detector or emitter, so that each photon that's produced is entangled with a spin qubit in a deterministic way. Quandela, a French company that is pursuing this strategy, claims that it gives them 100,000 times more efficient operations than the traditional probabilistic method for generating entanglement1. Whenever you talk about 100,000x — five orders of magnitude — that's the kind of breakthrough that can only really be powered by a novel scientific approach, replacing the prior approach rather than just iteratively engineering it. It's very hard to get one order of magnitude improvement when you're approaching something as an engineering challenge.

In many ways, we're still on the precipice in 2025 that we stood on in 2019. Qubit counts are about the same. The capabilities are still experimental. The applications are still hypothetical. There aren't many new algorithms — certainly none to match the elegance and theoretical solidity of Shor's, Grover's or the Harrow–Hassidim–Lloyd algorithm. The inability to declare victory in 2026, more than six years since the Google Supremacy Experiment, is causing some people to run out of patience and declare quantum technology a failure, but that judgment merely betrays a lack of perspective on the level of quantum's maturity relative to classical information technologies.

The timescale for quantum is not that of RISC versus CISC, or 5G versus 4G — all incremental improvements that still took years and years to develop and deploy. Instead, it's best to think of quantum as being a reset to the dawn of the digital age, of the classical digital age, in the mid-40s. And by that measure, we're somewhere before the invention of the monolithic integrated circuit — a comprehensive approach to robust scalability, a technological platform on which all of CMOS and all of the digital age has been built. We may even be prior to the invention of the transistor, because of the uncertainty around which modality of qubit is the best to pursue.

Everyone's convinced their qubit is the best qubit. Take superconducting, one of the oldest operating modalities tracing its origins to the end of the 20th century. With that history, one might be tempted to think the best practices for the modality have been discovered and a general consensus has emerged for how to build superconducting qubits. Instead, there is an increasing diversity in the "flavors" of superconducting qubits, including those optimized for using "cat states" to bias against certain types of errors, or GKP codes to bias against other types of errors, or are physically designed to do QLDPC efficiently. These have some resemblance to one another and some resemblance to the origins of superconducting qubits and transmons, but enormous variability.

This same pattern is true across all modalities. Trapped ions now have at least three or four distinct designs emerging and strategies for scale being explored. Photonics have the kind of variation between Quandela's approach or PsiQuantum's approach or Xanadu's approach or

Orca's approach. And it's really not clear whether we're going to see convergence on a single modality, a single design within a modality, some kind of hybridization of techniques and science from one modality merged into another — things like the bosonic codes of GKP being merged into the fermionic, quasi-particle operations in a superconducting system, for example — or if we're going to engineer very complex systems that combine multiple qubit types for different roles: for storage, for computation, maybe even for particular algorithms that have different attributes where gate speed is more important than computation duration or vice versa.

Systems engineering is emerging as a major accelerant to progress overall. John Martinis attributes his own success with the Google Sycamore chip to his systems thinking — not just focusing on the physics experiment of the Josephson junction itself, but every component of the vertically integrated system from submitting a task to executing and getting a result. This is much more of a computer engineering mindset than a physics mindset, and is characteristic of the ongoing friendly tension and interplay between the physicists and the engineers, the business people and the technologists.

Systems engineering is driving innovation globally — in companies like Emergence Quantum in Australia, which is working on CMOS control systems that operate within the cryostat at the four Kelvin range or below, or Quantum Machines, which is focused on making a more and more capable control stack at room temperature that integrates with things like Nvidia's DGX for researching and prototyping error correction in real time, with physical qubits and GPUs working in a tightly coupled fashion.

There's also an emerging category of system simulation which supports systems engineering thinking. There's a category of classical software called EDA, electronic design automation, which is critical to the design of today's enormously complex microprocessors with hundreds of billions of transistors. Those systems are modeled at scale in very large computational workloads and simulated down to a very fine-grained degree in electromagnetic simulations, so that designers can ensure that when they actually pull the trigger on the fab line, the chip that comes out does the things they want it to do. Now there's emerging a new category of EDA — a kind of quantum EDA, if you will. This starts with roots in the experimental physics lab, where toolkits like QuTiP and others have been used for years to simulate many-body systems when an experiment is being designed. But it is now maturing into something that can emulate a full physical qubit with all of its supporting and ancillary systems — a full logical qubit with all of its physical underpinnings, a full vertical stack from qubit up to room temperature control electronics. You can imagine the complexity of every step of that chain: from the many-body or two-level system of the qubit itself, to the many-body system of the logical qubit, to the interaction with the filters and amplifiers along the microwave control line, the electromagnetic dynamics, and finally at the control electronics the translation from digital to analog and analog to digital. Companies like Quantum Elements and others are exploring this from a startup perspective, and of course the giants of the existing classical EDA field, like Cadence and Synopsys, are watching the space very carefully and conducting their own research.

On the investment front, mirroring the surge in public market action, private investment from venture capital and others is robust, more than doubling from 2024 to 2025. There are now a number of dedicated quantum funds. The pioneers like France's Quantonation are now being joined by QAI Ventures in Switzerland, 55 North in Denmark, and IconQ in the US, among others — a very healthy sign, because one of the risks in this early stage has been unsophisticated venture investment in very early stage startups which can, with eyes squinted, look like a typical SaaS or hardware startup with a slightly longer runway, but in fact are a category unto their own which you can't really understand the dynamics of unless you have a broader understanding of this unique set of technologies and the progress toward commercial value.

And perhaps the most encouraging sign can be seen when we look back at the original declaration of quantum technologies being a priority at a national level for the United States — the National Quantum Initiative, signed into law in 2020, set to be reauthorized in 2025, a date that has continued to slip through machinations of a political nature, not anything specific about the commitment to quantum. It was followed very closely by the Quantum DELTA effort in the Netherlands we've mentioned before, and then one nation after another over the ensuing years announcing national quantum strategies, which Qureca now lists at 30-plus countries and $70 billion in public sector funding pledged.

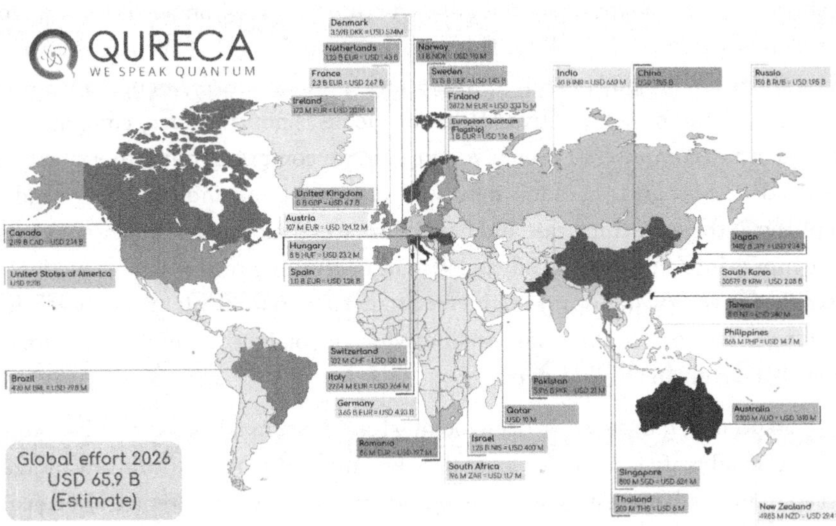

Figure 15.2: National quantum initiatives as estimated by Qureca

A healthy signal in and of itself, but even more encouraging is the development in the United States of increased focus at the regional level, at the state level, sometimes collaborative across state lines. These efforts are modeled after the Chicago Quantum Exchange, founded in 2018 by David Awschalom and Kate Timmerman of UChicago, with participation from UW-Madison, Northwestern, and UIUC — a remarkable piece of foresight on their part. They were, as the saying goes, skating to where the puck was going to be. And two years later, when the NQI Act was signed into law, CQE in Illinois landed two of the DOE NQI centers — SQMS at Fermi and Q-NEXT at Argonne — and one of the NSF centers, HQAN at UIUC. They mobilized a tremendous amount of activity around that initial effort, culminating in 2025 with breaking ground on the Illinois Quantum and Microelectronics Park, a 125-acre development on the former site of the US Steel plant — with all the attendant ironies and symbolism — and PsiQuantum and IBM Quantum as its two anchor tenants.

That vibrant ecosystem has attracted the attention of other states, with an acceleration of existing efforts between Colorado and New Mexico with Elevate Quantum, formation or refocusing of other existing affiliations like the Mid-Atlantic Quantum Alliance out of Maryland, giving rise to Quantum Capital — a billion-dollar budgeted effort by Governor Wes Moore to make Maryland more of a vital part of the quantum ecosystem, led by Corey Stambaugh, formerly of the OSTP and NIST. And net new efforts like Quantum Connecticut, where despite having Yale and UConn, there has not been the same kind of vibrant ecosystem.

What do we mean by ecosystem? This isn't just a pipeline to produce full quantum stack hardware vendors like Quantum Circuits in Connecticut or QuEra in Illinois, but instead a broader effort to create the foundational infrastructure that a region needs to begin to become quantum. This is a much broader lens. This involves something that Dana Anderson has said on the New Quantum Era podcast: "In 10 years, any company that's not operating at the quantum limit will not be competitive." What he means is not that every company has to be building a quantum computer or even using a quantum computer, but rather that every company needs to be cognizant of this new wave of engineering and technology and manufacturing at the quantum scale and how it impacts their business. That may be simply innovative inputs into a traditional advanced manufacturing process—construction businesses that have new quantum-engineered materials that are more energy efficient, lighter, stronger, cheaper, or more sustainable, which if you ignore, your competitors will take advantage of and gain an edge over you.

From this perspective, maybe we've been looking for quantum advantage in the wrong places. Maybe quantum advantage is much closer than we think. Maybe quantum advantage isn't a quantum computer solving a problem that would take a supercomputer a trillion years to solve — although that is still a possibility. Maybe quantum advantage is what's required to evolve our engineering and manufacturing processes past the two nanometer limit that defines ASML and TSMC's secret weapon in the semiconductor industry. Below two nanometers, you are effectively at a quantum scale, whether you like it or not. And so if you were designing denser microprocessors, they would have to be quantum in nature.

From this perspective, efforts like the IQMP, like Quantum CT, like Quantum Capital, like Elevate Quantum, are truly economic development efforts, broadly drawn. The study for the IQMP project estimated that 30% of the jobs that would be created wouldn't require a college degree. There's a job creation story there that's appealing to local politicians for obvious reasons. It's also an opportunity for a do-over. The CHIPS and Science Act had as one of its major objectives bringing the microprocessor industry back onshore after having offshored it willy-nilly and lost control of the value creation engine now dominated by ASML and TSMC in the Netherlands and Taiwan. It's very difficult to get a horse back in a barn once the barn door is open. However, quantum engineering and manufacturing represents a new foal — unsteady on its legs, unable to stand for very long, wobbly in its gait, unable to carry a load whether a rider or a pack, but full of potential and needing care and feeding in a barn, shelter, training, human stewardship, and investment.

It may be that the vision of millions of qubits operating on long-running calculations with the precision and fault tolerance required to, for example, crack a 4,096-bit RSA key or optimize a global supply chain — or do anything faster than a classical computer — is an unattainable vision. It's still possible that that is the case. However, broadly speaking, two outcomes with higher certainty are sufficient to justify the level of investment and attention that quantum technologies are getting.

First are the economic development opportunities of this new wave of advanced quantum engineering, which will affect every industry — directly, indirectly, or obliquely. The only unknowns are rate and pace. Some would argue that we're starting to see material design influenced by quantum technology today, with quantum sensing being applied on industrial manufacturing lines for levels of precision in detecting faults that were undreamed of before.

The other reason has to do again with what Deutsch calls explanatory power. The theory of quantum information, whether or not it ever led to workable technologies, has created a greater understanding of our universe. It has solved the black hole information paradox. It has inspired new efforts to explain the boundary between gravity and the quantum realm. It has inspired new theories of consciousness as an emergent property of the interaction of our neurotransmitters with the zero-point field that exists everywhere around us. And it will continue to unlock the mysteries of our universe.

When asked whether he saw himself as a technologist attempting to build quantum computing or an experimentalist attempting to probe the realm of the physical universe, Steve Girvin at Yale responded, "I'm an experimentalist. I just need large-scale many-body systems that are controllable in my lab, and for that I need industrial engineering, precision, repeatability, and durability to achieve those scales and that level of control."

It's very difficult to define what this point in time means precisely and what is on the horizon in any foreseeable scope of the future. But the macro and micro characteristics I've been describing paint an optimistic picture. In practical terms, nearly every commercial quantum hardware vendor building a full-stack system has a public roadmap that explains their strategy for fault tolerance and scale in their future R&D plans and milestones. And every one of them hopes to deliver a system to the market by 2030 that will be impossible to simulate classically.

There is a scale just slightly beyond where we are right now — something like 50 fault-tolerant qubits with full connectivity — that could create a computational space that is simply impossible for a classical computer to simulate. In one sense, everything from that point on that runs on that machine will be quantum advantage, because there won't be an alternative way to carry out those computations that take full advantage of the Hilbert space of 2^{50}. We don't know practically what you can do with 50 fault-tolerant qubits, partly because you can't simulate it, and partly because theory doesn't necessarily move in an incremental way.

Shor made a giant leap forward with the factoring algorithm, imagining something that would require some large number of physical qubits to run. He didn't then work backward to something that could run on five noisy qubits in 2000. The normal way that software and algorithms evolve is actually from a user of a device needing to scratch an itch — coming up with an innovative technique that fits the capabilities and characteristics of that particular device, running with it, expanding on it, talking about it with peers, who adopt and adapt it to new hardware and new use cases.

My favorite example of this: the mathematician Stan Ulam, at Princeton's Institute for Advanced Study, a collaborator with von Neumann, wanted to use the IAS computer for physics calculations having to do with neutron diffusion. While convalescing in the hospital after a bout of encephalitis and playing hour after hour of solitaire — because he was told that to help his healing process he had to think less — Ulam came up with a random sampling method that resembled some aspects of games of chance based on cards. Hence his name for it: the Monte Carlo algorithm. That was in 1948. The first use of the Monte Carlo algorithm to optimize a financial portfolio was sometime in the 1970s. Almost 30 years went by where that technique was known to a small but growing group of physicists and mathematicians as a useful tool, long before someone on Wall Street thought they might be able to squeeze some more return out of an asset portfolio by applying what is now considered one of the powerhouse algorithms of information theory.

It seems to me very likely that when that machine is in the market that can't be simulated — high energy physicists at CERN, material scientists at Fermi, chemists at PNNL, mathematicians at RIKEN, computer scientists at LRZ, and other scientists around the world will play with that device as a new tool, a unique portal into quantum many-body system dynamics at a scale previously impossible to manage and access. New science will come from those interactions, and that new science has intrinsic value unto itself — and will lead to greater and greater explanatory powers that will then translate into algorithms and technologies and applications and economic value and a better world for us all. The enormous potential of quantum information technologies remains; in fact, if anything, the estimates of value to be created have expanded. As a truly disruptive technology, however, the road to realizing that value will be long, and will require discipline, patience, imagination, and determination at a global scale, along a timeline which is impossible to predict.

You may think this is a cause for concern, but ask yourself, if you had lived in the 1950s and were told by someone who worked on von Neumann's machine at the Institute for Advanced Studies that one day that work would lead to an utter revolution that would reshape every aspect of human life across the entire world, would you have believed him? Would you have demanded to know what the use cases for this new information technology were? Would you have been intrigued, but impatient after a few years, to see this fabled computerized land of tomorrow? Truly revolutionary technologies demand a secular kind of faith in our human ability to keep plugging away, trying to connect imagination to intuition, intuition to scientific knowledge, knowledge to engineering, and finally engineering to a useful tool. Eventually we manage to find a way to recapitulate our mythology in technology, and I find no reason to doubt this will happen again with quantum technologies.

Coda: which qubit wins?

The two most commonly asked questions about quantum computing are: What are the use cases? — which I think we've just discussed — and which qubit is going to win? Predicting the future is a fool's game, however, if I had to hazard a guess, I would say that neutral atoms and trapped ions are both well-positioned in the near term. It is very likely that one or more vendors with atom-based qubits will ship machines before 2030 that are going to be seen as very valuable, possibly the first quantum advantage machines that are commercially useful. They have enormous plasticity with their all-to-all connectivity and the customization of the topology. Vuletić and Lukin's work on the fault-tolerant qubit demonstration on neutral atoms in 2024 essentially built a von Neumann machine out of atoms in a vacuum chamber, with storage and operations being partitioned and strict rules governing the movement of qubits from one space to another. That simply is not possible when you're talking about silicon or superconducting or even really photonic qubits — they all rely on more rigid topologies and architectures. That's an enormous power of atom-based quantum computing. While ion traps are not as flexible as the free space in a Rydberg atom vacuum, their all-to-all connectivity and very high fidelities make them extremely powerful as well.

However, atoms are very, very slow. It's possible that through parallelization techniques, there will be gains of 10, 20, 30, maybe a hundred times. But gate speeds on superconducting devices or spin qubits are much, much faster. And so sizable chemistry simulations, for example, while computable on a neutral atom device or a trapped ion device with a thousand qubits in a way that you can't do classically, might take six months or a year to complete. That's a lot of time for something to go wrong. A lot of time to wait for an answer. It certainly doesn't sound like something with an easy path to realizing commercial value. There certainly will be scientific value. And the plasticity of the platform will help enormously accelerate the design of other types of qubits by providing a test bench for error correction and topologies.

Superconducting qubits have the edge on speed — orders of magnitude more rapid than atom-based gate operations. However, they have significant challenges as well, primarily of connectivity, and scale. Whatever connectivity exists on the chip is the connectivity you get. Google's chips have been a lattice with each qubit in the center connected to four neighbors, and the edge qubits having only three or two connections. IBM's gains in qubit counts in the early 2020s came as a result of the heavy hex topology, which further reduced the connectivity between qubits. Theoretically you can still have two qubits on any part of the chip interact with one another, but you have to do swap operations which are costly and introduce overhead and greater challenges for error correction.

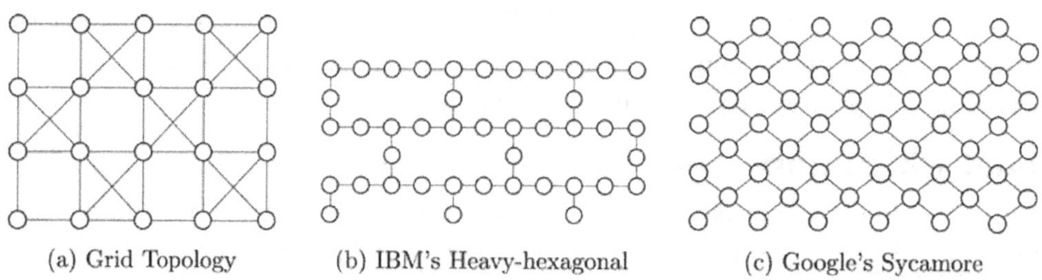

Figure 15.3: Hour, Leanghok & Heng, Sengthai & Heng, Sovanmonynuth & Go, Myeongseong & Han, Youngsun. (2024). Context-aware coupler reconfiguration for tunable coupler-based superconducting quantum computers. Quantum Science and Technology. 10. 10.1088/2058-9565/ad8510.

IBM is in the process of pivoting to a new topology, co-designed with their chosen error correction strategy based on QLDPC, which requires long-distance connections — that's the "LD." They're making a biplanar chip with 144 qubits on each plane connected so that certain qubits have long-distance connections used for error correction. That error correction only protects the Clifford gates, so the non-Clifford operations have to be done on a separate chip, a lattice that distills magic states and teleports gates into the biplanar chip. All of which seems admirable and remarkable, but also a measure of how much contortion is required to get a superconducting device to carry out error correction and harness those fast gate speeds in a way that is fault-tolerant and not massive overhead. Google's lattice, of course, is only suitable for surface code, which at a decent degree of fault tolerance — distance 7 or 9 — is something like a hundred physical qubits to one logical qubit. That's reduced to 12-to-1 at distance 7 in

IBM's scheme, but this is all an illustration of how hard it is to make superconducting qubits robust.

And when it comes to scale, the current known fabrication techniques are probably going to top out at maybe 10,000 or 100,000 qubits per dilution refrigerator. The problem is the microwave signaling leads — at the moment, every qubit needs several of those leads, which take physical space, bring electromagnetic field distortion or interference potential, and bring thermodynamic load to the fridge. So, in order to get to a million qubits, you may have to link ten dilution refrigerators — see the discussion of transduction above.

Qubit modality	Gate speed	Connectivity	Coherence	Scaling outlook	Leading vendors
Superco-nducting	Very fast	Layout-limited	Short–moderate	Fridge/ control bottlenecks	IBM, Google, Rigetti, IQM, Alice & Bob
Trapped ion	Slow	Near all-to-all	Excellent	Modular traps/ networking	IonQ, Quantinuum, Oxford Ionics, AQT
Neutral atom	Slow–moderate	Reconfigurable	Good–excellent	High raw counts	QuEra, Atom Computing, Pasqal, Infleqtion
Photonic	Potentially fast	Network-native	Loss-limited	Million-qubit ambition	PsiQuantum, Xanadu, Quandela, ORCA
Spin / silicon	Fast	Local/dense	Good–excellent	CMOS-scale promise	Quantum Motion, Diraq, SQC, Intel, Equal1

Table 15.1: ***Qubit modalities as competing engineering bets.*** *No modality has won; each trades speed, connectivity, coherence, and manufacturability differently.*

It is my belief that superconducting qubits will overtake atom-based qubits based on the knowledge that atom-based qubits unlock for the superconducting community. This is illustrated by Vuletić and Lukin's LDPC experiment influencing the community overall and, in some way, leading to IBM's choice of QLDPC—but then topping out at some amount of scale.

The scale question comes down to two leading contenders: spin qubits and photonic qubits. Certainly, PsiQuantum feels that they're on a path to delivering a million fault-tolerant qubits. That remains to be seen. Certainly, Quantum Motion, Diraq, SQC, and others believe that they're on a path to unlocking the technology of CMOS as a means of producing qubits. And if they do that, it should be said, there is no doubt that CMOS is the most successful human technology for producing scale you could ever imagine — there are trillions of transistors fabricated per second on the planet, as we've said. But the challenges there are significant and full of question marks for both. I don't think it's clear whether photonics or silicon will be first, but I believe that one of them will take the throne from superconducting qubits and deliver higher-scale devices. No matter which one wins, it is unlikely that that will be the end of the race. There is no point at which this age of heroic qubits gives birth to one perfect qubit that ends this story.

References

1. `https://www.quandela.com/about-us/newsroom/quandela-announces-a-100000-reduction-in-the-number-of-components-needed-for-fault-tolerant-calculations-a-major-breakthrough-for-photonic-quantum-computing/`

Subscribe to Deep Engineering

Join thousands of developers and architects who want to understand how software is changing, deepen their expertise, and build systems that last.

Deep Engineering is a weekly expert-led newsletter for experienced practitioners, featuring original analysis, technical interviews, and curated insights on architecture, system design, and modern programming practice.

Scan the QR or visit the link to subscribe for free.

`https://packt.link/deep-engineering-newsletter`

16

Unlock Your Exclusive Benefits

Your copy of this book includes the following exclusive benefits:

DRM-Free PDF Version

Download DRM-free PDF and ePub copies of this book.

7-Day Packt Library Access

Get 7-day unlimited access to 8,000+ books and videos. No credit card required.

Available for first-time Packt+ trial users only.

Next-Gen Reader Access

Read this book on Packt Reader with progress sync, dark mode and note-taking.

Follow the guide below to unlock them. The process takes only a few minutes and needs to be completed once.

Unlock this Book's Free Benefits in 3 Easy Steps

Step 1

Keep your purchase invoice ready for *Step 3*. If you have a physical copy, scan it using your phone and save it as a PDF, JPG, or PNG.

For more help on finding your invoice, visit `https://www.packtpub.com/en-us/unlock?step=1`.

> Note
>
> **Note:** If you bought this book directly from Packt, no invoice is required. After *Step 2*, you can access your exclusive content right away.

Step 2

Scan the QR code or go to `packtpub.com/unlock`.

On the page that opens (similar to *Figure 16.1* on desktop), search for this book by name and select the correct edition.

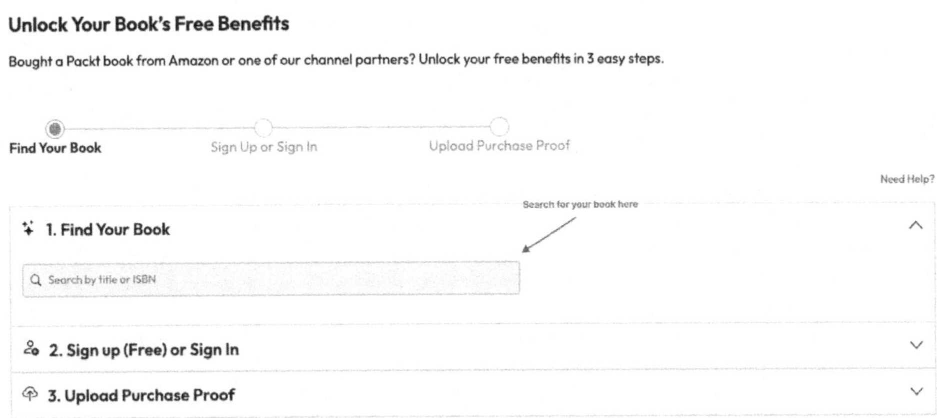

Figure 16.1: Packt unlock landing page on desktop

Step 3

After selecting your book, sign in to your Packt account or create one for free. Then upload your invoice (PDF, PNG, or JPG, up to 10 MB). Follow the on-screen instructions to finish the process.

Need Help

If you get stuck and need help, visit `https://www.packtpub.com/unlock-benefits/help` for a detailed FAQ on how to find your invoices and more. This QR code will take you to the help page.

> Note
>
> **Note:** If you are still facing issues, reach out to `customercare@packt.com`.

packtpub.com

Subscribe to our online digital library for full access to over 7,000 books and videos, as well as industry leading tools to help you plan your personal development and advance your career. For more information, please visit our website.

Why subscribe?

- Spend less time learning and more time coding with practical eBooks and Videos from over 4,000 industry professionals
- Improve your learning with Skill Plans built especially for you
- Get a free eBook or video every month
- Fully searchable for easy access to vital information
- Copy and paste, print, and bookmark content

At `www.packtpub.com`, you can also read a collection of free technical articles, sign up for a range of free newsletters, and receive exclusive discounts and offers on Packt books and eBooks.

Other Books You May Enjoy

If you enjoyed this book, you may be interested in these other books by Packt:

Learn Quantum Computing with Python and IBM Quantum

Robert Loredo

ISBN: 978-1-80324-480-8

- Get familiar with the features within the IBM Quantum Platform
- Create and visualize quantum gates and circuits
- Operate quantum gates on qubits using the IBM Quantum Composer
- Install and run your quantum circuits on an IBM Quantum computer
- Discover Qiskit and its many features such as the Qiskit Runtime
- Get to grips with fundamental quantum algorithms and error mitigation techniques to help you get started
- Understand the new era of quantum utility and how this moves us closer towards quantum advantage

Dancing with Qubits

Robert S. Sutor

ISBN: 978-1-83763-675-4

- Explore the mathematical foundations of quantum computing
- Discover the complex, mind-bending concepts that underpin quantum systems
- Understand the key ideas behind classical and quantum computing
- Refresh and extend your grasp of essential mathematics, computing, and quantum theory
- Examine a detailed overview of qubits and quantum circuits
- Dive into quantum algorithms such as Grover's search, Deutsch-Jozsa, Simon's, and Shor's
- Explore the main applications of quantum computing in the fields of scientific computing, AI, and elsewhere

Packt is searching for authors like you

If you're interested in becoming an author for Packt, please visit authors.packt.com and apply today. We have worked with thousands of developers and tech professionals, just like you, to help them share their insight with the global tech community. You can make a general application, apply for a specific hot topic that we are recruiting an author for, or submit your own idea.

Share your thoughts

Now you've finished *The New Quantum Era*, we'd love to hear your thoughts! Scan the QR code below to go straight to the Amazon review page for this book and share your feedback or leave a review on the site that you purchased it from.

https://packt.link/r/1807787370

Your review is important to us and the tech community and will help us make sure we're delivering excellent quality content.

Index

S

T

U

W

Made in the USA
Monee, IL
16 September 2026

63035035R00122